Limits of Scientific Inquiry

Essays by

David Baltimore
Sissela Bok
Harvey Brooks
Barbara J. Culliton
Loren R. Graham
Gerald Holton
Peter Barton Hutt
Leo Marx
Walter P. Metzger
Robert S. Morison
Dorothy Nelkin
Don K. Price
Robert L. Sinsheimer
Judith P. Swazey
Lynn White, Jr.

Limits of Scientific Inquiry

Edited by GERALD HOLTON and ROBERT S. MORISON

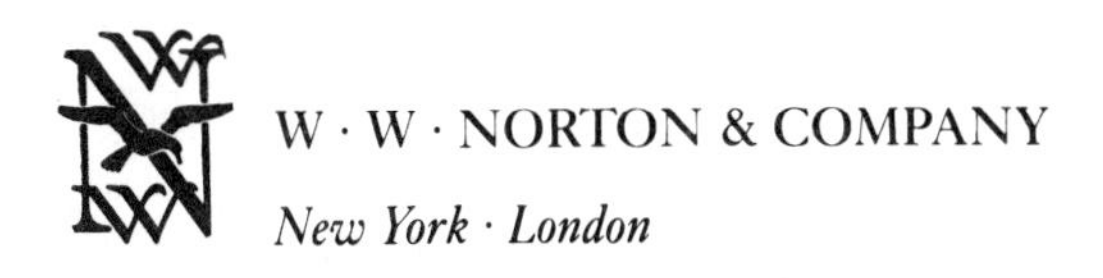

W · W · NORTON & COMPANY
New York · London

Library of Congress Cataloging in Publication Data
Main entry under title:
Limits of scientific inquiry.
First published in 1978 as v. 107, no. 2 of the Proceedings of the American Academy of Arts and Sciences.
Includes index.
1. Science. 2. Research. 3. Ethics I. Holton, Gerald James. II. Morison, Robert S., 1906–
Q158.5.L55 1979 507'.2 79–5103
ISBN 0–393–01212–3
ISBN 0–393–95056–5 pbk.

2 3 4 5 6 7 8 9 0

Contents

Preface

A SERIOUS VOLUME WITH the title of *Limits of Scientific Inquiry* would have been difficult to conceive of even a few years ago. To be sure, some special aspects have long been given attention, for example the safeguards on experimentation with human subjects. But the intellectual climate that prevails today, and the new awareness of risks that accompany certain technological advances, have thrown into prominence a whole range of ethical, political, and ideological issues. These involve scientists, engineers, and other professionals at least as much as the proverbial layman-citizen.

What we have sought to provide here is a balanced inquiry into representative portions of the whole spectrum of ideas and specialty fields. Most of the authors met as a group periodically for about half a year, and most of the papers in the volume were developed from the presentations made at these meetings.* The seminars, held at the Massachusetts Institute of Technology in its new College for Science, Technology and Society, were first called together by the Provost of M.I.T., Walter Rosenblith. He set the tone for the enterprise in his letter of invitation, where he described the circumstances that made such a study seem appropriate; he wrote:

> Some of these are practical and timely, such as the current discussion concerning the conditions for research on recombinant DNA or on human subjects. Others have their roots in basic, long-term, sociological or epistemological changes. Thus it is now maintained by many that scientists and scholars have long had a bargain with society by which they have produced ideas and devices with few constraints, but that now this bargain is in danger of breaking down or in need of revision. Understanding the extent and reasons for such changes should be of interest in its own right, and may also improve our ability to deal with practical problems some of us are now facing.

With the exception of the final essay, these contributions were recently issued as a volume of *Daedalus,* the quarterly of the American Academy of Arts and Sciences. The editor of the Journal, Dr. Stephen R. Graubard, helped us immensely in the planning and execution of our task. We worked with him in close collaboration as many of the essays were guided through several drafts.

*The Advisory Group for this volume, consisting of the core members of the seminar series plus a number of others who could come only once or a few times—some from distant cities—included the following: David Baltimore, William A. Blanpied, Sissela Bok, Harvey Brooks, Carlos Chagas, Stephan Chorover, Barbara J. Culliton, Norman Dahl, Michael Dertouzos, John Deutch, Martin Deutsch, Irven DeVore, John Edsall, Loren R. Graham, Michael Hammer, Harold Hanham, Hermann Haus, Gerald Holton, Peter Barton Hutt, Merton Kahne, Carl Kaysen, Kenneth Keniston, Leo Marx, Walter P. Metzger, Alexander Morin, Elting Morison, Robert S. Morison, Dorothy Nelkin, Frank Press, Alexander Rich, Walter A. Rosenblith, Jack P. Ruina, Vivien Shelanski, Robert L. Sinsheimer, Eugene Skolnikoff, Krister Stendahl, Ann Strauss, Judith P. Swazey, Lester C. Thurow, Stephen Toulmin, Sherry Turkle, Charles Weiner, Joseph Weizenbaum, Jerome B. Wiesner, and Joel Yellin.

From the beginning it was a joint effort, to assure that this first collection of studies on the general topic of the necessity and perils of constraints on research achieve the highest possible level.

G.H.
R.S.M.

Acknowledgments. This volume has been made possible through grants from the National Science Foundation Office of the Public Understanding of Science (grant No. OSS74–14718), the Commonwealth Fund, and the National Endowment for the Humanities. The Rockefeller Foundation, at a critical moment, assisted us with a supplementary grant. We are very grateful for the support, and for additional help and participation from the M.I.T. administration and from Dr. Graubard's staff at *Daedalus*.

The opinions, findings, or conclusions expressed in this publication are of course those of the authors and do not necessarily reflect the views of the sponsoring agencies.

ROBERT S. MORISON

Introduction

IN THE LAST FEW YEARS, several different kinds of unease have led to a questioning of the status of new knowledge and the effectiveness of society's arrangements for encouraging or restraining the growth of knowledge. In practice, the controversy changes rapidly from one level to another, and it is often difficult to be sure of the particular concerns and motivations of the principal protagonists. For purposes of preliminary analysis, however, several sources for this new anxiety may usefully be distinguished. The most elementary, perhaps, is a concern for the harm that may be done to individuals in the simple pursuit of knowledge. Closely related is the concern for the possible damaging effects of new technologies that may result from new knowledge. Next come the long-term hazards, hard to foresee except rather dimly, that carry some finite possibility of serious perturbations in our current way of doing things. This category might include more remote results of genetic engineering, culminating possibly in the development of "a whole new race of human beings," or the climatological changes that might possibly result from altering the transparency or heat capacity of the atmosphere. The fourth source of anxiety is somewhat more remote from everyday affairs and concerns the possibly unsettling effects of new knowledge on man's concept of himself and his relations to society or the rest of the natural world. This last concern may take the form of a deep-seated and not always clearly verbalized anxiety about the possible limitations and bases of scientific knowledge itself. The recent counterculture literature has provided insistent critics of these epistemological inadequacies, but the theme has a venerable and significant history, as the essay by Leo Marx in this issue reminds us.

Finally it may be worth noting that unease in regard to science reflects a generalized decline in public esteem for authority figures of all kinds. Scientists are thus suspect, not only as authorities in their own right, but also because the scientific establishment has become identified with the general power structure, which to some of our citizens, at least, appears to become ever more overbearing and untrustworthy. We may be witnessing here the result of an exchange of roles between science and religion in relation to the stability of the prevailing political system. Many, if not most, of the nineteenth-century anarchists and revolutionaries were intensely secular if not actually atheistic. Thus they tended to embrace science as an ally in the battle against governmental authorities who continued to rely heavily on an established church to defend the legitimacy of a temporal power still remembering its divine origin. Today's social activists, on the contrary, see science as occupying somewhat the same positions that religion used to in support of an authoritarian state. Some indeed have come full circle and use revelations and theological arguments to weaken the theoretical base of modern political power. Some of these originate from far outside the scientific community, as when the Creationists attack the National Science Foundation for putting the authority of government funding behind a secular presentation of man. But even within the scientific community there are attitudes that remind one more of quasitheological than strictly scientific considerations. Most striking perhaps is the concern of scientists like Robert Sinsheimer or Erwin Chargaff about tampering with man's genetic constitution or crossing the "natural" barrier between pro- and eukaryotic DNA.

In the real world, the various themes roughly identified above are more often than not intertwined with one another. Furthermore, the protagonists on both sides may often be implicitly motivated at one level while they draw their arguments from another, where the issues may be simpler to present and more convincing to the assumed audience. It therefore has seemed useful to begin our collection with a detailed analysis of public concerns about the effects of science and technology. Loren Graham draws on his knowledge not only of the contemporary scene but of the historical development of science to present a more finely grained and complete taxonomy than the one just sketched. It also has the merit of suggesting the points at which external efforts to direct or limit the development of new knowledge might be most and least appropriate. Perhaps the most important function of such classification is to provide the various parties with a frame to which they may refer to make sure that they are talking about the same thing at the same time.

In the increasingly public debates about science policy, the most audible emphasis has recently been put on the need for restraints. Even more upsetting to those most affected by the imposition of such restraints is what seems to be a rather sudden shift of emphasis taking place in the United States, away from internal regulation or self-regulation, and toward external governmental interference. Not the least of the novelties is the fact that research of presumably universal interest should find itself at the mercy of citizen groups, from Princeton to Ann Arbor to Berkeley. But the difference of opinion does not simply pit town versus gown, scientist against nonscientist. There are deep differences within the scientific community itself. On the one side are those—a minority, ranging from earnest and eloquent scientists of the old style to a tiny but shrill radical cell—who would, at the very least, agree with the remark in the report of the committee on scientific freedom and responsibility of the American Association for the Advancement of Science, entitled "Scientific Freedom and Responsibility," that "some of the consequences of science and technology frighten, discourage and even horrify most of us."

Among the more distinguished and respected spokesmen for this view, especially as it concerns the issue of DNA research, have been Erwin Chargaff, of Columbia University, and Robert Sinsheimer, of the California Institute of Technology. The latter has been developing his views on these matters for the last two years, and the article he has prepared for us represents the current state of his thinking. Although deeply personal in tone, the paper reflects the unspoken, often unformulated feelings of what is at present a relatively small, but significantly sensitive, fraction of the scientific community.

As Sinsheimer's is possibly the best classical statement of a position on one side, David Baltimore's essay fulfills the analogous function on the other side. He reflects the current state of thinking within this tradition in his straightforward and confident paper. Many of his colleagues have expressed themselves more strongly and with less appreciation of what is troubling the other side. For example, the president of the National Academy of Sciences was recently quoted as having said, "Particularly troublesome is the ever more frequent expression of the notion that there are questions that should not be asked, that there are fields of research that should be eschewed because mankind cannot live with the answers. Nonsense. No such decision can be rational, much less acceptable." Or consider James Watson's characteristically pithy remark: "All I can say is that we want to go on with what we are doing, and I don't think we are crazy." Nonscientists too have warned scientists against accepting externally imposed limits to inquiry.[1]

Having pointed out the spread of opinions on the current state of affairs, it is well to pause and ask how we got where we are today. Most histories of modern science begin with the Renaissance in Western Europe, with Francis Bacon talking science like a lord chancellor, and Galileo rolling bronze balls down inclined planes (or, in another version, simply thinking about doing so). But as Lynn White tells us in his fascinating and imaginative paper, the roots both of the desacralized view modern science has about nature, and our painful ambivalence about scientific findings, lie much deeper, in the Middle Ages, when men drastically revised their notions about God, the nature of nature, and the criteria for proof. For those twentieth-century students who hope to escape history, it may be sobering to learn that the current tendency of those with high scores in quantitative studies to dismiss the literary arts as "trivial" is simply a distant echo from the masters of the medieval quadrivium similarly derisive of the merely verbal studies of the trivium.

One wishes there were more space and opportunity for contemplating the continuities in the long struggle to gain some congruence between the working of man's brain and the operation of the world around him, but we do have time to join Leo Marx, who in his essay takes us on a tour through the nineteenth century, and the romantic reaction against a metaphysics that ascribed an increasing degree of reality to numbers as it pushed man's need for wholeness further and further into the background, finally to bury sweetness and light in the dark satanic mills. The nineteenth-century romantics are known to us primarily through the "moderate" wing, with such representatives as Wordsworth, Emerson, and Carlyle, who struggled to maintain a place for scientific understanding as they listened with perhaps a greater yearning for the impulse from a vernal wood. Their twentieth-century counterparts, as Professor Marx explains, have abandoned the struggle for complementarity, with its tolerance for uncomfortable ambiguities. They offer us in exchange "the whole picture," a product obtained by the largely undefined procedures included under the curiously antique heading of Gnosis.

The critics may too easily overlook the fact that the metaphysical inadequacy of science is in one sense simply the price it pays for its wide acceptance. Whatever else it may be, science is necessarily a social consensual activity, and its vision of the world, precisely because it is limited, is one that can be widely shared. Indeed it *must* be shared, since its very validity resides in the reproducibility, under appropriate circumstances, of the evidence upon which any new conclusion depends. Gnosis, for all its subjective wholeness, must, like other purely subjective experiences, exact its price in terms of the basically private or individual character of the world it reveals. Thus the antisocial, anarchic character of the counterculture is not simply a coincidence. It stems precisely from its metaphysics.

Even if the problems between science and society were traceable to the metaphysical bases of science, it is not these, but the secondary attributes and practical results of science, that are in the forefront of today's discussion. In tracing the developments of the last twenty-five years, Don Price's essay quite properly sticks closely to these practical matters. Much of the predicament of science in relation to society is, as he shows us, traceable to the great change in the status of science, as an activity, in relation to its social and political context. In a very real sense science has exchanged its birthright for a mess of pottage. Only a few can still remember how things were when the soup plates were often empty, and scientists depended for project support on a few crumbs from rich men's tables. The war, and the portentous appearance of a Sputnik in the sky lifted science off to unprecedented financial heights at a price that is only now becoming clear.

The birthright, like perhaps all birthrights, is not without its ambiguities. "Scientific freedom," or "the right to inquire," are commonly-heard phrases, but what they describe is not easy to discover. At the 1976 annual meeting of the American Association for the Advancement of Science, Harold Green of George Washington University Law School explained in plainest terms why experimental scientific inquiry cannot expect blanket protection under the First Amendment, although he did not rule out all hope for those who would pursue "pure" knowledge with only paper and pencil.[2] Now Walter Metzger traces the development of conventional notions of academic freedom during the early nineteenth century in Germany and later in the United States and shows us how little they offer in the way of a general defense for freedom of scientific investigation in its twentieth-century form.

Thus the scientific community may feel itself left somewhat naked to its enemies so far as matters of principle are concerned. This is especially painful, perhaps, because those who would restrain or redirect the scientific effort claim to have at hand a compelling body of philosophical reflection, ethical principle, and legal precedent. In fact, the field of ethics, as an academic discipline which had steadily declined during one hundred years of secularization of our universities, has suddenly burst into flower with the pent-up excitement of a century plant. Why has this flowering come about? Largely, it seems, because the growth of science and its associated technologies has forced upon us questions of high ethical moment. This resurgence of moral dialogue has, on the one hand, shaken the scientist's confidence in the immunities supposedly granted to him on the basis of his claims of value-free objectivity. On the other, it has helped restore the confidence of those brought up in other disciplines who have always felt uneasy about science without being able to do or say much about it.

Sissela Bok, who carries one of the steadiest lamps through these still somewhat obscure corridors, summarizes for us the current thinking of the scholar in the bioethical field on the rights and wrongs of scientific experimentation, especially as it directly affects human beings. Judith Swazey has spent her professional career observing the behavior of physicians in ethically ambiguous situations, and she tells us why it is unlikely that even the healing professions, much less the traditional, "distinterested, pure scientists," can be left to regulate themselves. On the other hand, her detailed exposition of the codes for clinical investigators, developed since the horrors bared at the Nuremberg trials, leaves one wondering how much of virtue can be effectively enforced through external, legal controls. But let us hasten to observe, as Barbara Culliton reminds us, that the situation we are describing is far from a struggle of bad guys on the inside and good guys outside of science. Although some physicians on occasion overenthusiastically pursued experiments that led them to inject malignant cells into uninformed elderly patients, others have been among the leaders in the current rebirth of medical ethics. Indeed it was the biologists themselves, rather than philosophers, politicians, or public representatives, who first directed attention at a possible need for restraint in the investigation of recombinant DNA. To be sure, the enthusiasm of some of the critics from within science does not stem only from an unalloyed concern for ethical principle. As in all evangelical movements, one sooner or later becomes aware of the complexity of human motivation, with its undertones of secondary psychological needs.

At all events, clearly defined principles will play only a small part in what several of the papers refer to as the renegotiation of a hitherto tacit concordat (to use Don Price's term) worked out decades ago between society and science. Instead, as we shall see in a few moments, the ultimate agreement will doubtless deal largely with practical matters. In all likelihood it will also remain partly unexpressed, even if there are written documents and standing commissions to herald the new order. It would certainly be a mistake, however, for the scientific community to conclude that "the public" is completely disillusioned with science and technology and determined to regulate them "like the railroads." Barbara Culliton reminds those few inclined to panic that there is no evidence that the man in the street has lost confidence in science. The more significant novelty is provided by the "men and women who have made a serious, essentially full-time career representing what they describe as the public interest." Like many crusaders, they have motives which are mixed and they tend to interpret certain aspects of scientific behavior in ways which are quite foreign to the scientists themselves. How different such views may be becomes more apparent as one reads the bill of particulars drawn up by Peter Hutt, one-time general counsel for the FDA and a student of interactions between government, industry, and science. He here assumes the robes of an avocatus diaboli to produce a list of criticisms of science as seen from a number of obtuse and occasionally acute angles. Those scientists who continue to hope that the recent preoccupation with the side effects of research on recombinant DNA is but a temporary perturbation might do well to read over the list from time to time to remind themselves that things really are very different from what they used to be even a few years ago.

Inspection of his list reveals a pervasive concern with the way priorities are set, or more simply, with how the scientific world decides what to do next. Scientists on the whole have believed that the best, perhaps the only way to decide what to do next, is to look at the scientific work itself. Normally there is an inner logic to what is going on in any given laboratory as well as in the field as a whole. It is this that gives science its accumulative character, and it is this same inner dynamics that has until recently guided the funding, at least of basic science, and often of applied science as well. Harvey Brooks explains in detail how these properties have been used in the past to guide science policy and to set priorities. His analysis also reveals some of the difficulties that arise when attempts are made to combine this classical mode of inner direction with an effort to respond to external pressures and social needs. The conventional wisdom among foundation officers and government officials has been, on the whole, that it is better to follow good *leads* than to try to satisfy felt *needs*. But perhaps the greatest successes have been scored when both lead and need coincide.

In any event the situation now is that the tax-paying public has been persuaded to supply for research unprecedentedly large (though as a fraction of the GNP still very small) amounts of money on the promise of good things to come. When the good things don't come fast enough, this same public, taught by recent events to question the motives of all its leaders, stands ready to question even those of the scientific mandarinate. The list Peter Hutt has been able to assemble is the unhappy result. The worst of it is that the implications may be at least half right. Many scientists have perhaps made too much of a point of their disinterest in practical results in an attempt to establish their position as humble seekers after truth. Too often they have simply succeeded in appearing as intellectual snobs following their own interest at public expense. A public which may have some difficulty appreciating truth for its own sake finds it easy to imagine less elevated, more personal motives.

Hutt's paper makes no pretense as an analytical essay or reasoned recommendation. It is simply a list of possible criticisms to be taken into account by a somnolent scientific community as it struggles to understand the changing context within which it must learn to live. One is reminded of the title of an earlier reforming chorale: "Wachet auf, ruft uns die Stimme."

Dorothy Nelkin has for some time studied the public dissatisfaction, if not with basic science, at least with its technological results. The effort to be heard on such matters as the siting of power plants, the distribution of medical care, science education, and now biological research, has given rise to a still rather confused search for a new machinery to implement the as yet insufficiently focused demands for participation and/or accountability. The actual design of such machinery must be high on the agenda of anyone concerned about the future of science and technology, but first let us look at what we have now.

Most scientists and an increasing number of nonscientists who have looked into the matter are quite happy, often indeed enthusiastic, about the system devised in the forties and fifties for judging the scientific merit of research proposals and arranging funding priorities on this basis. On the other hand there is far less public or professional satisfaction with the means at hand for appraising *social* merit. Contrary to what might be supposed, the problem of social relevance was recognized, at least in principle, from the very beginning. For that reason the membership of the National Science Board of the National Science Foundation explicitly included men and women of affairs as well as working scientists. More significantly, perhaps, the councils attached to each of the National Institutes of Health include lay representatives of the public and are specifically charged with the reviewing of all proposals from the policy point of view. This is not the place to go into detail, but my overall impression is that these policy-making and monitoring bodies have not yet found how to function as satisfactorily as those lower in the echelon charged with the determination of scientific merit only. One symptom of inadequacy has been the need to appoint special ad hoc commissions when policy questions of particular social interest arise.[3]

Any attempt to redesign the social and political machinery for developing funding priorities for science and technology more closely in accord with social need might well begin with a careful examination of why what we have now has not done a better job. Curiously, several investigations have been made of the peer review mechanism for determining scientific merit while the admittedly less satisfactory procedures for determining social relevance remain relatively unexamined. It is not the primary purpose of this collection of papers to suggest specific administrative and political change, but rather to lay the groundwork for further thought. Nevertheless the closing pages of the papers by Loren Graham and Dorothy Nelkin should be useful for their presentation of the specific social needs that any new system of science policy planning must meet.

My own essay elects a more direct, pragmatic approach and asks how any conceivable policy-making machinery might deal with some of the current proposals to restrict the growth of knowledge. By no means all potentially "dangerous" research is promoted by the scientific community in opposition to the wishes of society. Sometimes, as in the case of research on aging processes chosen for special examination, the public, or its representatives, has taken the lead. Thus I tend to feel that in the real world, decisions are likely to be made on the basis of majority preferences which in turn are based on what can be more or less easily foreseen or guessed about the practical results of proposed research. In the four actual cases considered it appears to be unnecessary to reach the theoretical or constitutional question of freedom for the investigator, for in all of them the solution that is likely to be chosen on practical grounds is also the one that does not restrain the investigator. Furthermore, at this writing, the agitation for a special commission to regulate research on recombinant DNA appears to be receding in the face of technical answers to the safety problem and growing doubts about the wisdom of interfering for more esoteric reasons.

Nevertheless, recurring fears about the implications of new knowledge remain real enough to justify further exploration of basic principles. Even if one agrees that in practice the "limits to inquiry" are likely to be defined more as a matter of legislative policy than of constitutional rights, it may be worth looking a little deeper for the root of the matter, if only to provide some theoretical background for the legislators who must define policy. Actually the need for some hypothetical "right" or other formal means of protecting freedom of inquiry arises only when the dangers are visible enough and the benefits doubtful enough to cause a majority of the people or an unusually vocal fraction to be against proceeding with the proposed inquiry. Almost everyone agrees that a right to inquire provides no defense for investigators who might wish to proceed in the face of such clear and present dangers as may be inherent in the process of investigation itself. But what about the dangers that may come about secondarily as a delayed result of the new knowledge uncovered by the inquiry? At this point it is hard to see why the scientific investigator should not enjoy equal rights with the logician or the religious mystic. The mere fact that he manipulates material in a laboratory or observes the stars through a telescope in addition to employing a pencil and paper does not justify suppressing an investigation directed at increasing the store of general knowledge (unless the experimental procedures are themselves dangerous).

Here the distinction between basic and applied or targeted knowledge becomes crucial. The point, of course, is that the more basic or general a bit of knowledge is, the more it may be put to several different uses, some benign and others undesirable. When this stage is reached, arguments such as those of Green and Jonas that the First Amendment does not protect the scientific investigator lose much of their force.[4] The responsibility for the adverse effects would seem to be more appropriately borne by those who actually use the knowledge to commit undesirable acts than by those remote creators of new knowledge who simply render the act theoretically possible.

As Loren Graham reminds us, however, some of the most intense hostility to scientific ideas is not occasioned by their results in terms of practical technology but by their conflict with other, presumably more agreeable, ideas about the nature of man and his place in nature. Astronomical theories that question the centrality of God's interest in humankind, or biological theories that propose close kinship with the lower animals have historically been opposed on the ground that they degrade the quality of man's spiritual life, or for the more pragmatic reason that they reduce the incentives for good behavior. Professor Sinsheimer's concern about the probability that successful interstellar communication would reveal beings better than ourselves would appear to lie in the first category. Much of the public objection to evolutionary theories of the descent of man seems to settle on the second. These differences in opinion seem to lie precisely in the area of political, social, and theological debate that the First Amendment was designed to protect. It might, of course, be argued that scientific modes of knowing are more dangerous and therefore more deserving of suppression than are religious revelation or philosophical speculation because of the greater degree of verifiability within their sphere of competence. But it scarcely seems proper to curtail the freedom of scientific thought simply because of its higher probability of being right.

What may be a more compelling variant of the preceding argument holds that scientific statements that turn out to be false are more dangerous than the false statements of other disciplines, because they enjoy some transfer of authority from statements of similar form and derivation that are known to be true. Although this contention may hold for short periods, it should not be troublesome over the long run. Statements made in the scientific form are by definition and intent potentially falsifiable in a way that statements derived from revelation and pure reason are not. Thus, if Jensen's conceptualization of group differences in IQ is wrong, it will be shown to be so by peaceful means. Intuitive or quasiphilosophical theories of the master race could only be overcome by force. Indeed, history shows that many more wars have been fought and more individuals burned at the stake because of differences over statements made as a result of revelation, intuition, and pure reason than as a result of scientific investigation.

A somewhat more subtle form of the foregoing antiscience argument is the contention discussed by Leo Marx, that the admitted triumph of the scientific method in what Descartes called the world of extension has made it very difficult to believe in any other kinds of knowledge or any other kind of world. As a result, human beings are denied access to the essential elements of their own humanity. Although Theodore Roszak regards this outcome as a disaster and actually recommends "repeal" of the scientific world view as a cure, it is not likely that society could restrain scientific investigation simply because of its metaphysical limitations; much less could it abolish memories of the past success that makes it so threatening.

In all humility, it must, of course, be admitted that it is impossible categorically to deny that we may have reached a point where we must abandon the faith that knowledge is better than ignorance. We simply lack the ability to make accurate predictions in cases such as this. By the same token, it would seem an act of extraordinary arrogance to assert that we *can* recognize the moment at which we should cast aside the most essentially human elements in our adaptive armamentarium.

It is exceedingly difficult to summarize a discussion that is being carried out simultaneously on so many different levels by so many different protagonists with so many different objects in view. One overall impression is that in spite of the evidence of declining faith in established ways of doing things, in spite of the time lost at the laboratory bench by those who are confronting matters of high policy for the first time, in spite even of the personal hostilities occasionally engendered, the current far-ranging, often rather random review of the place of science in society has been and will continue to be a good thing. The scientific community has led a peculiarly unexamined life for a surprisingly long time, and many have accepted its unusual and, until recently, unquestioned status a little too easily. Indeed, in the last twenty-five years, in an effort to raise its financial support at a rate nearly triple that of the rest of society, the scientific community may have promised too much too soon. Certainly it underestimated the demand for accountability that sooner or later extends to every region of our democracy.

In the long run it will doubtless be good for scientists to have their consciousnesses raised on these matters. However, one wonders if, in the heat of the moment, the emphasis has not been a bit misplaced. Granting that the social and political machinery for maximizing the benefits of science and minimizing its hazards is in need of considerable improvement, is it really necessary or even desirable to begin with talking about setting limits to inquiry or regulating laboratories the way we do the railroads? Would it not be more sensible, and in the long run more productive of good, to ask how science can be helped to become more beneficial to the public than how it can be kept from being more harmful? The distinction is much more than a purely semantic one. Society is almost certainly in greater danger from what it *does not know*—for example, about the stability of ecological systems, about the physical chemistry of the stratosphere, about the growth of human population—than from what it *does know* about inserting small snippets of mammalian DNA into bacteria or even cloning new strains of sheep.

The fact is that we have not found good ways of encouraging certain kinds of much-needed inquiry, especially perhaps in the areas of the environment, the control of population growth, and the conversion of energy that loom so importantly in our future. In none of these areas does the primary difficulty appear to lie in the lack of appropriate methodologies or conceptual frameworks. The method and the intellectual leads are there, at least for a better beginning than has been made. The problems are of a different, more embarrassing kind. Much of the needed work is boring and lacking in prestige. In some cases years may be required before publishable results can be expected. Often the work requires a high degree of cooperation between many different individuals and disciplines. Last but not least, the findings may make difficulties for influential vested interests in industry, or even in government itself. All these factors combine to make many of the needed studies uncongenial to the prospective investigator and poorly adapted to traditional ways of doing things in universities and granting agencies. The same is true of much of the meticulous monitoring and assessment needed to predict and control the undesirable effects of technology, an effort that must occupy a far higher place on the agenda than it has in the past.

If the growing public interest in science policy can be matched by a growing awareness of the public interest on the part of the scientific community, and the two together can find a better way of ordering the priorities for science and technology, the present somewhat inchoate discussion will have been well worthwhile.

References

[1]Gerald Piel, "Inquiring into Inquiry," *Hastings Center Report*, 6 (August, 1976): 18.

[2]Harold P. Green, "The Boundaries of Scientific Freedom," *Newsletter on Science, Technology, and Human Values* (June, 1977): 17.

[3]For example, the National Commission for the Protection of Human Subjects of Behavioral and Biomedical Research. Besides considering the general problem, it reported on such special matters as the use of fetal material for research purposes and the ethics of psychosurgery. The President's Biomedical Research Panel established on January 29, 1975, spent fifteen months conducting "an extensive study that involved assessments of the state of the science, the impact of federally funded research on institutions of higher education . . . the dissemination and application of research findings, and the development of policy for federal support of biomedical and behavioral research."

[4]Cf. Green, *op. cit.*, and Hans Jonas, "Freedom of Scientific Inquiry and the Public Interest," *Hastings Center Report*, 6 (August, 1976): 18.

LOREN R. GRAHAM

Concerns about Science and Attempts to Regulate Inquiry[1]

CONCERNS AND ANXIETIES about science and technology are not novel in the history of science, but a persuasive case can be made that a new stage has been reached in the last decade or so, a stage which has been marked not only by growing suspicion of science among segments of the lay public, but also by an expanding belief among scientists themselves that questions about the social responsibilities of researchers must be faced more directly than has previously been the case. The best publicized example of this awareness has probably been the controversy over recombinant DNA.

Other challenges to the unfettered development of science and technology have appeared in a great variety of areas, including biomedical research, military research, and nuclear energy. As a result of all these controversies, some scientists have voiced their worry that "scientists and scholars have long had a bargain with society by which they have produced ideas and devices with few constraints, but that now this bargain is in danger of breaking down or in need of revision."[2] In the spring of 1977, at an MIT discussion of this new resistance to science and technology, one participant spoke favorably of proposals to "regulate science as one regulates the railroads," while others met this concept with sharp criticism.

It will be the thesis of this article that discussions of proposals to regulate science do not have general value until a number of basic distinctions have been made among types of concern about science and types of proposal for its regulation. What we need as a first step toward a discussion of possible limits to inquiry is a typology or taxonomy of concerns about science. We can then assess the validity of each concern and address the problem of limits or regulation in a more specific and informed fashion.

The categories which I have devised are based upon two different sources of information: my own perception of the recent literature about the effects of science and technology on American society, and the results of a survey sponsored by the National Science Foundation several years ago entitled "Attitudes of the U.S. Public Toward Science and Technology."[3] The major element has been my own interpretation, but I have tried to construct a system of categories which incorporates most of the answers given in the NSF survey to a question inquiring what "harmful things," if any, come to society from science and technology. Near the end of this essay I will raise the question of how well the taxonomy of concerns presented here fits the results of that public opinion poll.

The main goal in classifying the concerns about science is not merely to introduce order, or to engage in a possibly interesting academic exercise. The goal is much more pragmatic and is, indeed, clearly policy-oriented: to sort out those categories of concern about science and technology where we are coming to recognize the necessity or inevitability of controls of one sort or another, while retaining other categories where freedom from control is still defensible and necessary. There is still a large category of research where regulation of science should not be permitted. This autonomy of science should be defended not as a privilege for an elite, nor as an absolute right, but as a need of society itself. The conclusion will be, then, that some aspects of science and technology should be controlled for the same reason that other aspects should not be: namely, it is better for society that way, society taken in the broadest terms of all of its interests.

A limitation of this analysis is that many concerns about science do not easily fit into a defined framework, because they are expressed in poorly defined or even irrational terms. Indeed, the most basic fear of science that might be discussed is probably the very fear of rationality itself. Nonetheless, I think that at the present time there is a particularly strong case for discussing separately the "rational variable" in the complex cluster of contemporary concerns about science. There are at least three reasons for such an approach. (1) By isolating the anxieties about science that can be phrased in rational terms, we may decrease the apparent dimensions of the irrational residue. (2) The rational criticisms of science need to be evaluated on their own ground. (3) Some of the concerns about science and technology are legitimate, in the sense that they represent genuine fears for the safety and security of society. If we dismiss all concerns about science as "irrational," we will not be listening to some important debates.

The subsequent discussion will be based on a number of broad categories of concerns about science and technology, some of which have subcategories:

I. Concerns about Technology	
A. Concerns about the physical results of technology	"Destructive Technology"
B. Concerns about the ethical results of technology	"Slippery Slope Technology"
1. Biomedical ethics	
C. Concerns about the economic results of technology	"Economically Exploitative Technology"
II. Concerns about Science	
A. Concerns about research on human subjects	"Human Subjects Research"
B. Concerns about distortions in allocations of resources for science	"Expensive Science"
C. Concerns about certain kinds of fundamental knowledge	
1. Knowledge itself	"Subversive Knowledge"
2. Knowledge "inevitably" leading to technology	"Inevitable Technology"
D. Concerns about accidents in the research itself	"Accidents in Science"

E. Concerns about the use of science to excite racial, sexual, or class prejudices	"Prejudicial Science"
F. Concerns about certain modes of knowing	"Ways of Knowing"

The list given above sorts the concerns into the two broad divisions, technology and science. Some critics will respond that this division is no longer tenable. In an eloquent article Hans Jonas maintained "not only have the boundaries between theory and practice become blurred, but . . . the two are now fused in the very heart of science itself, so that the ancient alibi of pure theory and with it the moral immunity it provided no longer holds."[4]

Although I appreciate the argument which Professor Jonas has advanced, for the following reasons I do not consider it a serious challenge to the classification I am presenting here. First, I do not maintain, as Jonas' comment seems to imply, that problems of "science" enjoy a "moral immunity." I will consider ethical and moral issues under the headings both of technology and of science. (Note, for example, "human subjects research.") Second, although I agree that there are some problems which cannot easily be classified as either science or technology, it is my opinion that the great majority of current issues can be so classified without much difficulty, including most of the ones which have recently attracted so much attention in the press. The division between fundamental studies of physical and biological nature, on the one hand, and technological studies directed toward social or economic goals, on the other, is still a useful one. Third, I will consider issues under "Inevitable Technology" where the boundary between theory and practice has become blurred in the way in which Jonas emphasizes.

Destructive Technology

Concern about the physical results of technology is one of the most familiar and easily identified categories of concern. A prime example is the damage to the environment caused by industrial civilization. A specific case would be damage to the ozone layer alleged to result from the escape to the upper atmosphere of fluorocarbons used in aerosols. Others would be the polluting effects of DDT, supersonic transports, and various energy sources. The large array of issues under this category also include many that are not best described as environmental, such as regulation of the distribution and use of pharmaceutical drugs, food additives and chemicals (like saccharin), explosives, and radioactive materials. All are examples of society's need to prevent the products of technology from spreading in an unmonitored or unregulated way.

This category is the one where the observation "if we can regulate the railroads, so can we regulate science" is most apt, although we should remind ourselves that we are speaking here of technology, not science. The imposition of controls on the use of destructive technology is clearly within the rights and traditional practices of government regulatory agencies. Important questions about the wisdom of the regulations, economic and otherwise, can be raised, but the principle of government regulation is well established. Whether the sale or use of aerosols with fluorocarbons should be controlled is a problem of technical facts and assessments, not one of principle about limitation on freedom of inquiry.

Some concerns about military technology also fall within this category. Individual scientists differ widely on whether they should engage in work on destructive weapons, but few people question the principle that governments have the right—indeed, the obligation—to regulate and control this activity. The original McMahon Act of 1946 placed strict controls on the ownership and production of fissionable materials, although it was much less restrictive (and more intelligent) than the May-Johnson Bill, which favored military rather than civilian control.[5]

Controls in this area can extend down to the level of the individual, as the McMahon Act did. If it is possible to discuss regulating hand guns, it is certainly possible to continue to control possession by private individuals of materials necessary for manufacturing nuclear weapons and even to prohibit unauthorized research on methods necessary for such manufacture, such as laser separation of isotopes.

Slippery Slope Technology

Whereas the previous category dealt with the physical effects of certain technologies, the present one centers upon the ethical effects. Obviously, every act of physical damage contains an ethical dimension, and no clean separation of the two can ever be made. Nonetheless, there exists a significant difference between, on the one hand, a concern whether the physically damaging effects of a certain technology can be excused within existing ethical systems, and, on the other hand, whether a certain technology may be destroying the ethical system itself. The present category deals with the latter concern.

The use of new technology in the biomedical area has recently raised many difficult ethical issues. New possibilities for prenatal diagnosis by amniocentesis, prolongation of life through the use of dialysis and respiratory machines, psychosurgery, and DNA therapy (or "genetic engineering") are only a few examples of issues which by now have been widely discussed in the press. A whole new field of discussion, biomedical ethics, has developed as a result of the advent of these new technologies. The Institute for the Study of Society, Ethics and the Life Sciences in Hastings-on-Hudson and the Kennedy Institute Center for Bioethics in Washington, D.C., are two of the more active centers among the many where issues of medical ethics are now under close study.

One of the major concerns expressed by observers of technological developments in the biomedical field has been that by blurring or erasing ethical boundaries that were earlier considered absolute, we will go out onto a "slippery slope" of relativistic ethics on which we may lose our balance and tumble to the bottom. Critics often raise the specter of eugenics and medical experimentation under national socialism as the ultimate point of descent. They ask, if we sanction the disconnection of life-support machines in terminal cases, or we perform abortions on genetically defective fetuses in the last months of pregnancy, what ethical limitations exist preventing us from taking a more active role in ending the lives of terminal patients suffering pain, killing a deformed child immediately after birth, or even several months after birth? The possibility of becoming callous to moral values through adoption of the radical proposals or repetition of the less radical ones is worrisome, and properly demands our attention. On the other hand, if we simply prohibit abortions or termination of extraordinary care, we will encourage worse ethical abuses (illegal abortions under frightful conditions, economic injustice, even infanticide by parents). The slippery slope slants in more than one direction.

Controversial as these issues are, they do not, in themselves, directly involve the question of limits to inquiry. They are immensely difficult questions of the wise use of medical technology. Guidelines are developing which help to prevent abuses while still taking advantage of the benefits of the new technologies, benefits which in many instances are significant and deeply humane.

Whether or not regulation in this area is justified is not in doubt (it *is* justified), but what the regulations should be and what person or institution should devise or enforce them raise fundamental problems. Current practices in regulation differ widely. For example, where chemicals might be used, as in proposals for DNA therapy (DNA is, after all, a chemical, and therefore subject to more rigorous controls than pure surgery, which is up to the doctor and the patient) the Food and Drug Administration in the United States will probably play a large role.[6] The rulings of the courts are already important in abortion and disconnection of life-support mechanisms. Regulation of technology falling under this category is already an accepted and necessary practice, albeit requiring enormous care.

Economically Exploitative Technology

Enormous costs are involved in research and development in certain high technology areas. Should large public sums be spent for the development of an American supersonic transport when only a small portion of the population would ever utilize such transportation? Should large amounts be invested in developing exotic and cosmetic medical treatments if these sums detract from public health programs? How does one decide how much money should be given research on diseases such as cancer and heart disease which afflict highly industrialized societies to a greater degree than underdeveloped societies (where, for example, kwashiorkor and schistosomiasis may be more serious threats)? From a health standpoint, how does one decide the relative importance of research on curative medical technologies as opposed to improvement of environmental conditions, which are increasingly seen as the source of many illnesses?

All the above examples refer to control of research and development for future technologies, not the deployment of existing technologies, and in that sense relate rather directly to the question of limits to inquiry. Yet the general public usually does not make this distinction. When the pollsters for the NSF survey asked citizens what, if anything, they feared about science and technology, they sometimes voiced an economic concern quite different from the "priorities problems" discussed in the previous paragraph. They said technology "puts people out of work by replacing them with machines" and "it has forced a lot of small businesses out of work." This concern is, of course, one of the oldest in the history of modern technology, and can easily be traced to the very beginnings of the Industrial Revolution. The fact that the overall result of industrial expansion has been greater employment with greater variety of tasks is little consolation to the individual worker made surplus by the disappearance of his or her craft or task.

A final example of economic results of technology is the distortion in the use of natural resources which certain technologies entail. The present effort to slow the exhaustion of fossil fuels by automobiles and home-heating is an example of control of expensive or wasteful technology.

In each of the examples discussed above, the principle of control for economic reasons is already established in the United States, although great controversies about the best means of control remain. In contrast to the centrally planned economies, where direct limitation on outputs of certain technologies is often imposed, in the United States the emphasis has frequently been on economic penalties or rewards (taxes, deductions) on the technology, and on controlling research and development through modulation of federal research grants. This approach seems quite proper. Still, the possibility of controls of a variety of types, particularly in response to crisis situations, is apparent. If these controls are carefully executed within a democratic framework, they are visualizable without damage to political or academic freedom.

As we move now from the subject of concerns about technology to that of concerns about science itself, we come much closer to questions of principle that are essential to our concept of free inquiry.

Human Subjects Research

The methods of procedure in certain types of research may do damage to human subjects of the research, or be suspected of doing such damage. The adoption of guidelines on research on human subjects by the National Institutes of Health is an example of efforts to avoid damaging effects of this type.

Surely we all admit that limits to inquiry *do* exist within this category. One cannot, for example, morally justify injecting human subjects with pathogenic organisms or toxic chemicals to determine their lethality. Nor can we determine the outer limits of human toleration of physical or psychological deprivation, or of physical pain, by direct experimentation, even if subjects would volunteer for such experiments. The history of medical experiments in Nazi Germany, such as the ones designed to help the Luftwaffe decide how long a downed pilot could survive in icy waters, is still a recent memory warning us against such immoral and crude experiments.

This category of concern about science is occasionally lumped together with "slippery slope technology" issues in the biomedical field, but a distinction between the two is useful in discussions of possible guidelines or regulations. The essential question already discussed in the category of slippery slope technology is the direct result to the individual patient and the indirect ethical results to society at large of the use of certain types of medical technology; the goal there is therapy, not acquisition of fundamental knowledge. The category presently being discussed is research involving human subjects in which the goal of therapy is either absent or secondary to the acquisition of information.

Since the main question here is of research design, the present category touches more closely than the earlier ones the problem of freedom of research. Some of the debates in this category have been quite heated (for example, the one on XYY chromosome research), with some scientists fearing the intrusion of public controls into fundamental research itself.

Separating the category of "human subjects research" from that of "slippery slope technology" reveals several interesting trends which have significance for policy recommendations. Usually we say that fundamental research is distant from moral and legal considerations while applied science is closely bound to such restraints. Yet a comparison of this type of fundamental research with medical practice in recent years reveals some interesting departures from this usual situation. The advent of life-support technologies has de facto given physicians greater and greater leeway in making life-and-death decisions, while in fundamental research on human subjects the ethical leeway granted to researchers has been progressively diminishing. Popular opinion has often supported this divergence. The physician who may have facilitated the early death, either through action or inaction, of a horribly suffering terminal cancer patient is usually not questioned closely about his or her practices. Often the close family members do not know exactly what the physician has done, and they do not wish to know. The situation may be one of unconscious complicity, with both physician and family desiring the early death. The law may have been broken in some strict sense, but a "higher morality" was supposedly served.

In research directed toward acquiring information rather than treating a particular patient, however, the ethical restraints are growing, not relaxing. Finding a cure for yellow fever in the way in which Walter Reed and his colleagues actually did it would now be considered ethically questionable, since the controlled experiments exposed human volunteers to the disease.

Despite the new restrictions, some of which have troublesome implications, most of us would probably agree that greater ethical awareness about research is a laudable development. Research scientists have not only a moral responsibility to avoid damage through individual and corporate examination of the possible effects of their research procedures on human subjects, but also an interest in avoiding such controversies by self-regulation, since the alternative is increasing regulation by bodies outside the scientific community. The fact that the law already provides some avenues for redress to individuals actually harmed is not a sufficient answer to the problem. The responsibility of scientists themselves for the alleviation of public concerns is heavier in this category of research than any other so far discussed.

Expensive Science

This is a category of concern about distortions in the allocations of resources for science. One of the most frequent observations made about contemporary science is that research in many areas is no longer possible on individual budgets, or even small institutional budgets, and that large external support is now necessary. Although external support of scientific research is centuries old in several countries, the degree to which such support is obligatory if high-level research is to be continued has changed markedly in this century. Robert Boyle, Antoine Lavoisier, Charles Lyell, and Charles Darwin were all outstanding scientists who were able to support themselves entirely or partly on the basis of their own funds. Now even a Nelson Rockefeller who was a radio astronomer would probably not be eager to undertake the building of a very large array radiotelescope.

As external support of science has grown, the degree of involvement of governmental organizations and large private foundations in making decisions about allocations of resources has increased in step. Congress has a proper and natural interest in the size of the budget of the National Science Foundation. The question of freedom of inquiry does not arise here in a pointed way so long as the external bodies are generally favorably disposed to science and allow qualified specialists to evaluate individual proposals. Possibilities for conflict involving academic freedom do exist, however, if the budget-granting bodies try to evaluate the quality of individual applicants. The attempted Bauman Amendment to the NSF authorization in 1975 would have permitted Congress to review individual NSF research grants on a systematic basis and would have been a serious threat to research autonomy.[7]

The concerns of fund administrators about scientific research impinge on freedom of inquiry in several indirect ways not yet mentioned. If a line of research has no visible practical value, the scientist in the field will probably have fewer possible sources for funds than if he or she can point to a potential valuable technology issuing from that research. If the research is highly unorthodox, it may not receive support even if potential applications are visualizable. Fund administrators traditionally fear controversy, and they fear even more the possibility that the work which they support will bring discredit to the funder (as did the Carnegie Foundation's support in the early decades of this century of the eugenics movement). An administrator of another large private foundation recently offered to show its early archives to an historian, but he said he did not have records on unsuccessful applicants, only on those who received grants. The records of the unsuccessful petitioners might be more interesting.

At the level of decision-making about the construction of the most expensive research installations, such as mammoth high-energy accelerators like the one at Fermi Lab, a worry about distortions in the allocation of resources is a serious one. The important discussions of societal priorities necessary for making such decisions do not present a fundamental limitation on individual freedom, even though in such areas work cannot be done without outside funds. Society obviously does not have an obligation to build a facility costing millions of dollars for every good scientist who wants one. On the other hand, a prosperous society which decided to stop building such facilities would be doing damage of a double nature: it would be inflicting harm on its scientific community and it would be limiting its own curiosity and creativity.

Subversive Knowledge

All the categories of concern about science and technology discussed so far involved side effects; they were based on the physical, ethical, or economic results of individual technologies, the effects of the design of fundamental research in projects involving human beings as subjects, and the economic costs of large fundamental research projects. The concern to be discussed now rests on fundamental knowledge itself.

Some of the best-known cases of interference with science in past centuries have been the results of resistance to fundamental knowledge. In the classic instances the anxiety has had two sources: the new fundamental knowledge was seen as conflicting with the theories of ruling authorities, and it appeared to demote the place of man in nature. The affair of Galileo's censure by the Church is one which can be interpreted in terms of these concerns. The Catholic Church had, undoubtedly without logical necessity, incorporated the basic features of the Ptolemaic system into its body of theological teachings. Furthermore, the rival Copernican theory seemed to demote the place of man by situating his abode outside the center of the Universe.

Historians of science will be quick to point out that the situation was far from being as simple as the foregoing description might seem to indicate. Copernicus dedicated his treatise to a Pope, and his work was not criticized by Rome until many years after his death; Protestant leaders such as Melanchthon and Calvin protested at first more vociferously than the Catholic leaders; the Copernican system was actually not heliocentric but heliostatic; the Catholic Church might not have taken the position it did were it not under the pressures of the Reformation and Counterreformation.

All these qualifications are valuable, but the fact still remains that on a general but nonetheless very significant level the Copernican system as espoused by Galileo was perceived as both contrary to theological teachings and demeaning to the place of man. It was a form of pure knowledge that was seen as a threat in itself. Debates over Darwinism in the last part of the nineteenth century also contained these two elements: conflict with at least some interpretations of religions, and an apparent denigration of the uniqueness of man.

We often assume that such conflicts are now elements of the past, that a mature world which attributes such an important place to science has surely outgrown adolescent anxieties of this type about science. And, indeed, we have improved very much in this regard. The concerns are still there, however, and they have appeared in several new forms. Furthermore, some of these concerns are not trivial.

Some of the popular resistance to studies in the field of ethology and primatology can be related to concerns about the diminution of the uniqueness of man, the narrowing of the distance between humans and the rest of the animal world. The possibility of increasingly successful explanations of human behavior in terms of animal behavior often evokes resistance among intellectuals who would usually consider themselves far from being antiscience. One does not have to accept the vulgarized interpretations of ethology of the type of Robert Ardrey or Desmond Morris to agree that at least part of the resistance here is the ancient one of concern for man's place in nature.

Another example of current concern about the uniqueness of humans is that expressed by the biophysicist Robert Sinsheimer elsewhere in this issue and earlier, at an Airlie House conference, where he commented,

> Should we attempt to contact presumed 'extraterrestrial intelligences'? I wonder if the authors of such experiments have ever considered the impact upon the human spirit if it should develop that there are other forms of life, to whom we are, for instance, as the chimpanzee is to us. Once it were realized someone already knew the answers to our questions, it seems to me, the impact upon science itself would be especially devastating. We know from our own history the shattering impact more advanced civilizations have upon the less advanced. In my view the human race has to make it on its own, for our own self-respect.[8]

This objection to possible research cannot be described, it seems to me, as anything else than another variation of anxiety about man's place in nature of the type often seen in past centuries, though not in a religious context.

On this topic of resistance to fundamental knowledge in itself my position is a simple one, and, I would maintain, of great importance to our intellectual freedom. That position is: "Critical discussion, yes! Regulation, no!"

In fairness to contemporary commentators such as Sinsheimer, who are responsible and helpful critics, I should add that many of them are not suggesting censorship or a ban on research in these areas, but a deep questioning of the amount of national resources that should be invested in certain research. But to ask whether millions of dollars should be directed toward possible contact with another civilization is to transform the question from one of limits to fundamental inquiry to one of allocation of resources, which is the separate category "expensive science."

Inevitable Technology

Another form of concern about fundamental science is the one that states "anything that *can* be done, *will* be done," and therefore, the critics say, it is not justifiable to draw a line between science and technology when one is attempting to discuss the question of limits to inquiry. These commentators continue that since some sorts of knowledge will "inevitably" lead to technology which will "inevitably" be used (even though we all might agree that the specified use is not desirable) we ought to impose limits on the original inquiry. Holders of this viewpoint might even classify certain types of knowledge as "forbidden" in principle.

Although I believe that this form of argument is incorrect as a general position, unnecessarily condemning our civilization to technological determinism, I would like to present it in the strongest form before criticizing it. We will all agree, I think, that fundamental physical research in the first four decades of this century provided the knowledge necessary for the construction of nuclear weapons; most of us would agree that the major political powers in the world now possess these weapons in sufficient numbers that if all were exploded in the atmosphere either our civilization would be destroyed or it would be so altered that many of us would not be eager to survive in whatever remained. If that holocaust should occur, the following argument would ring hollow in the ears of anyone able to hear it: "Science itself has no moral responsibility for what happened, since it was not necessary that nuclear weapons be constructed or used, even though the knowledge necessary for these events was produced by science."

Logically correct though the argument may be, we are all familiar with the failure of logic to handle adequately the most extreme human dilemmas. The simple fact would be that without the development of nuclear physics this ultimate disaster would not have occurred.

But on the basis of this realization should we try to regulate fundamental research in the future in order to avoid such events? I believe that we would not know how to do so even if we wished, and that in the effort we would cause immense damage to our existing values as well as deprive ourselves of many benefits, both material and intellectual. Frightening as our present situation is, it may just force us to be more responsible than we have ever before been.

The example of nuclear science may be misleading for investigations of "inevitable technology" in those less apocalyptic areas where we are trying to make decisions about the relationship of science to technology. Usually we have a little more freedom of maneuver, a little more space in which to make a few mistakes, than we do in the area of nuclear weapons. To a much greater degree than ever before we are adjusting to the need for controlling our technology instead of letting it control us, and therefore decreasing the strength of the link of necessity between science and technology. In areas of transportation, the environment, energy use, and city and town planning, we are trying to bend technology to our social needs. If we can succeed in this effort, the argument that "whatever can be done, will be done" diminishes in intensity. Society obviously needs some things to be done much more than others.

Taken literally, the "inevitable technology" argument is clearly false. Examples can be given of available technologies which have never been employed on a wide scale. As Harvey Brooks has reminded us, "Artificial insemination among humans is a good example of a technology which, though still in limited use, has never 'taken off.' People prefer the conventional method, and probably always will, when it works."[9]

The alternative of controlling fundamental research instead of technology is illusory, because it assumes the impossible: the foreseeing of the results of fundamental inquiry. Controlling technology is extremely difficult, but it is not impossible.

Accidents in Science

The controversy over recombinant DNA is one which needs to be analyzed on several different levels. The first level is the one which addresses the explicit and most common concern expressed by people who have objected to this research, namely that as a result of an accident in fundamental research the public safety will be threatened. This concern is the one which I have classified as a separate category, "accidents in science." The second level is one which concerns those largely unspoken, but nonetheless real concerns which relate to other categories. I will treat them separately.

On the first level, it is important to notice that the opponents of recombinant DNA research in Cambridge, Massachusetts, were not (at least originally) objecting principally to what could be intentionally produced by a successful recombinant DNA technology, but to the possibility during fundamental research of the accidental production of a pathogenic organism immune to normal antibiotics. It is true that one *could* express concerns of the type of category I ("technology") about recombinant DNA work (e.g., its possible use for biological warfare, or an objection to what might intentionally be done to alter the life of plants, animals, or people) but this type of resistance has so far not been the major one. The Cambridge City Council, for example, explicitly excluded these latter concerns from its considerations.

The topic of accidents in research is not a new one; indeed, accidents of various types, ranging enormously in the scale of potential damage, have long been evident. Explosions occur even in undergraduate chemistry laboratories, and electrocution is often a possibility. Even some common chemical elements, such as mercury or sodium, must be handled with care, and the dangers increase with volatile and explosive chemical compounds, radioactive materials and high-energy experiments. The transportation, distribution, and use of the most hazardous of these materials are already under controls of various types. In biology, unwanted organisms may escape to cause subsequent damage; the release of the gypsy moth in New England many decades ago continues to disturb us today.

One can include under this same category the possible inadvertent results of an introduction, for scientific reasons, of foreign materials into a previously undisturbed milieu. One example is the West Ford tests, in which in 1963 a belt of fine copper needles was placed in orbit 2,000 miles above the earth in order to investigate the influence on radiomagnetic propagation patterns. The project was criticized by other scientists because of possible side effects, and at least one report of such effects was published in the literature.[10] The same type of concern could possibly enter into seismological research in which water under high pressure might be pumped into strata of the earth in earthquake zones. Incurring responsibility for "causing" an earthquake may not be just a speculative possibility.

The second level of analysis which needs to be applied to recombinant DNA research goes far beyond the issue of public safety. As the controversy has progressed, the objections to this type of research have broadened beyond the original issue to a whole series of concerns that belong outside the category "accidents in science." Jeremy Rifkin has said, "The central issue is the mystery of life itself. It is now only a matter of time until scientists will be able to create new strains of plants, even alter human life." Rifkin's concern is a combination of several types: "destructive technology," "slippery slope technology," and probably a deep fear of science itself, one with nonrational roots. Ethan Singer has commented that DNA research "will eventually tinker with the gene pool of humanity. So the public, like the subject of any experiment, must give its informed consent—but willingly, not by coercion." This statement is a mixture of concerns about "inevitable technology" and "human subjects research." Roger Noll has added his view that since the federal government sponsors much of the research, "I suppose the question, 'Is this worth buying?' is going to be the real issue." This opinion represents a concern for "expensive research," but it is not clear whether Mr. Noll considered it science or technology, and it is furthermore not clear whether the financial concern is a cover for other worries.[11]

It is this mixture of concerns of various categories that made the recombinant DNA case so inflammatory. Any one of the concerns, by itself, could probably be resolved. Some of the anxieties are legitimate, although still not defined (e.g., public safety, which falls under my category "accidents in science"), while others are highly dubious or unacceptable to the scientific community (e.g., concern for the "mystery of life," which falls under my category "subversive knowledge.")

An important step in trying to deal with the concerns expressed about recombinant DNA is the attempt to sort out and evaluate them one by one. This step cannot be accomplished without first dividing "concerns about science" from "concerns about technology," the two most general categories. Part of the difficulty of discussing recombinant DNA research is the interweaving of these concerns, which in large part results from the fact that fundamental research in recombinant DNA is based on a novel "technique"—the introduction into a bacterial cell of a segment of foreign DNA by means of a now familiar multistep procedure—and therefore seems to be technology. Yet the development and use of novel techniques for conducting fundamental research is an old feature in the history of science (e.g., mass spectrometry in chemistry), and in no way defines the work as technology. It is true that the distinction is more complex with recombinant DNA than is usually the case, because the technique which is important to the fundamental research is also, apparently, the main method by which technology will probably proceed; nonetheless, the difference between exploring the nature of living organisms and intentionally trying to produce ones with specific socially useful characteristics is still important.

The separation of recombinant DNA fundamental research from recombinant DNA technology may be somewhat difficult to perform, but it is my opinion that the distinction should be made wherever possible. The policy implications are considerable. In the case of science, the emphasis should be upon keeping the work as free from controls as is consonant with public safety; in the case of technology, the emphasis should be upon the question of whether the application of the science serves a social need. The greater the risk of possible damage, the greater the social need should be before the application is developed and employed in society.

If one concentrates on fundamental research itself, the only concern about recombinant DNA which commands our attention persuasively is that of accidents which might endanger the public. Adequate protection against such accidents now seems feasible, but whatever the outcome of decisions in this area, it seems rather clear that the recombinant DNA case is not typical of many other areas of fundamental research. It would be a great mistake if recombinant DNA were taken as a paradigm case for regulating fundamental research. There have been relatively few actual cases in the recent history of science in which the primary concern expressed by knowledgeable critics was not about an application of science, nor financial, nor an ethical issue in biomedical research, but instead a worry that during the process of fundamental research an accident would result in social damage. The closest parallel to the recombinant DNA case is probably concern about accidents in nuclear research (e.g., the controversy over the Triga reactor at Columbia University), not to be confused with accidents in nuclear power installations (a "destructive technology" fear). The category "accidents in science" will probably broaden in coming years, but it is difficult to conceive that it would include more than a small portion of all fundamental research.

The recombinant DNA controversy is significant in two other senses: (1) the authentic issues it raises about who should bear the responsibility of devising and enforcing regulations on research, and (2) the possibility that underneath the specific concerns expressed by the public on recombinant DNA lurk much more significant and deeper irrational fears of science. These two issues are extremely difficult ones to resolve (more will be said on the first in the Conclusions), but little progress can be made in handling them unless distinctions have been made about various types of concerns about research, as I have been attempting to do here.

Prejudicial Science

A type of concern which is separable from those discussed so far is that scientists will present evidence or arguments that exacerbate racial, ethnic, sexual or class prejudices and which might be used in the service of a particular ideology. The old controversies over eugenics and the newer ones over intelligence tests and race can be related to this category, as can, to a certain extent, the current one over sociobiology.

There are governments today which ban such research on principle. The current edition of the East German encyclopedia *Meyers Neues Lexikon* contains in its article on "Race theory" (*Rassentheorie*) a condemnation of the subject and then the following statement:

> Race theory is still today openly propagandized in the imperialistic countries and especially in West Germany. In the German Democratic Republic the spreading of this doctrine is constitutionally prohibited.[12]

Research on intelligence and race would not be supported in East Germany. Neither is freedom of speech.

So far as formal regulation is concerned, my position on this category of concern is the same as on "knowledge itself": Critical discussion, yes! Regulation, no! I would not be candid, however, if I did not admit that I consider this category of concern to be one in which a pure separation of facts and values will probably never be possible, and may not even be desirable. If I were at present an executive of a research foundation and I were asked to vote on funding a program for projects on race and intelligence, I would vote negatively. I would refer to the fact that current definitions of "intelligence" and "race" are inadequate for the task, that the problem is not one that seems currently solvable, and that money could be used to greater advantage in other areas. And yet I know, and would acknowledge, that my negative opinion on the matter was influenced by my own values, my belief that the more potentially dangerous to society the results of research might be, the more rigorous one should insist that the methodology for that research must be.

If other people have the same opinion, and if the personal views of fund administrators at foundations and elsewhere are influential on this topic ("We are not going to do any work in this area"), is this not, in effect, the denial of freedom of inquiry in an area of fundamental research? In my opinion, freedom of inquiry is not foreclosed so long as: (1) no regulations are made prohibiting the research or the publication of the results; (2) fund administrators do not prohibit giving money for such research within "general category fellowships and grants," even if they refuse to support special projects of this type; (3) administrators do not interfere with the conduct of such research in institutions.

The problems presented by this category are immense, and I do not pretend to have solved them here. In the final analysis, the tasks of preserving both freedom of inquiry and social justice in this area depend on the existence of a nonpolarized society. A social environment so hostile to research of this type that no work could be done would pose, in fact, a true limit to inquiry of a sort that would be a dangerous precedent; on the other hand, a social environment in which certain political groups eagerly seized and successfully exploited arguments linking intelligence and race would present an extreme threat to society of another sort. So far we have managed to avoid both perils fairly well, and there is reason to hope for the continuation of this situation.

Ways of Knowing

A final type of concern about science stems not so much from any piece of information that science might produce, nor from any technology that might result from it, but from a critique of its mode of cognition. There are many people who maintain that science is only one of several avenues to knowledge, and these people often resist the tendency of science to expand its claims. The critique of science which these people advance attacks the epistemological bases of science on the ground that it is, at best, so specialized that it misses the most significant modes of reality, and, at worst, fundamentally alienating to the human spirit. The supporters of this view, who vary greatly in their sophistication, often call for supplementing the scientific approach with "other ways of knowing."

A contemporary example of such a critic is Theodore Roszak, who observed in his popular *Where the Wasteland Ends* that "Science is not, in my view, merely *another* subject for discussion, it is *the* subject." He spoke of the "intimate link between the search for epistemological objectivity and the psychology of alienation: that is, to idolatrous consciousness." "It is no mere coincidence," he continued, "that this devouring sense of alienation from nature and one's fellow man—and from one's own essential self—becomes the endemic anguish of advanced industrial societies."[13]

One should not assume that all proponents of "other ways of knowing" come from outside the scientific community. The distinguished astrophysicist and mathematician Arthur Eddington surprised and irritated some of his scientific colleagues by writing in the late nineteen twenties and early thirties popular books about science (which were adopted in many universities) in which he spoke of a world of "intimate knowledge" not accessible by scientific methods. Eddington wrote that when we make use of the "eye of the soul" as an avenue of cognition we should dispel the feeling that "we are doing something irrational and disobeying the leading of truth which as scientists we are pledged to serve."[14]

Such viewpoints have always existed in our history and undoubtedly always will. Whether one finds them valuable or not is a deeply personal characteristic. The views, fears, and premonitions of Goethe and Blake are only prominent examples of the many that could be associated with this category.

There is at least one area where this category of concern about science has current practical significance and where the term "regulation" may arise in at least some peoples' minds. In California and elsewhere, proponents of "creationism" criticizing Darwinian evolution have had some political success in their efforts to revise school textbooks by making the argument that since religious interpretations of the universe and scientific interpretations are based on strikingly different or even incommensurable ways of knowing, the contrasting interpretations should be given equal time. The creationists say that since neither they nor the evolutionists can disprove the views of the other side, both sides should be represented in high school and elementary science textbooks. This specious argument has an appeal to some listeners because of its superficial resemblance to the principle of giving all sides in a true debate opportunity for expressing their opinions.[15]

It is conceivable that this threat to science could become a serious one; the controversies over the MACOS (Man: A Course of Study) project and the resulting attempts to restrict NSF's autonomy are recent events in which this type of concern played an important role.[16] At the present moment, however, the strength of "alternative modes of cognition" may be receding with the passing of the peak of interest in mysticism and the occult in the late sixties or early seventies. At any rate, a form of special regulation that goes beyond the well-established separation of church and state does not seem appropriate. On the whole, school boards are more sophisticated than they were in the days of the Scopes trial, and one can only hope that the efforts to oppose religious, romantic, or mystical viewpoints to those of science will not receive significant official support.

Before drawing conclusions I would like to mention the NSF public opinion poll I drew on in attempting to classify concerns about science. The results of that poll may be misleading, because responses to the question "You've said that science and technology have done more *harm* than good—Can you tell me *one* of the harmful things?" came from only *2 percent* of all respondents, those who had already responded that science and technology have done more harm than good. Although the 2 percent may be a very thin and probably unrepresentative slice of American opinion, their responses do provide empirical evidence on an issue where such evidence is very rare. (Since the results of the NSF poll were only of limited value, my attempt to classify concerns about science and technology in a way that would be useful in approaching the issue of regulating research drew primarily on a more helpful source of information, discussion in journals and newspapers about the effects, real or alleged, of science and technology.)

Quite a few of the responses can be related to one or more of the categories of concerns I have sketched. (See reference 17 for some verbatim responses.) Most of the responses were directed toward technology, not science, with far more falling into "destructive technology" than any other category. Some responses are too indefinite to permit categorization. Others refer to superstitions or irrational fears. Few are sophisticated or reflective. No criticisms at all were voiced about the possible use of science to reinforce racial prejudices, none about "accidents in research," and none about recombinant DNA since the poll was taken before that controversy began.

Conclusions

The first major conclusion which issues from the analysis presented in this article is that there exists a core of categories of concerns about research where regulation should still not be permitted at all. All of them fall under the heading science, and not technology, but not all categories under science are in the core. This core consists of concerns I have labeled "subversive knowledge," "inevitable technology," "prejudicial science," and "ways of knowing." The most inviolable category of all is "subversive knowledge," for this is the area where the truly fundamental threats to science have emerged in past centuries, and where there is always the possibility of new dangers.

In the other core categories a democratic and healthy society should be able to avoid controls, but I will admit that I can imagine extreme and unlikely situations in which a temporary control of "inevitable technology" might be conceivable (a bit of fundamental knowledge which led easily to a serious and uncontrollable destructive application), and perhaps even "prejudicial science" (a bit of fundamental knowledge that would exacerbate social conflict in a temporary and extremely destructive fashion), but our society is presently far from being in such situations. (The circumstances would have to be roughly equivalent to the crying of "fire" in a crowded theater; for a hypothetical example, see reference 18). And in the category "ways of knowing" we might possibly need additional regulation in extremis to *protect* science, not limit it (which is what the principle of separation of church and state already does).

I am not willing, however, to admit that controls of any kind are justified on "subversive knowledge" under circumstances not covered by one of the other categories. It is my opinion, furthermore, that in the three other categories just mentioned ("inevitable technology," "prejudicial science," and "ways of knowing") there is no justification at the present time for regulation of research, and also that the likelihood of the emergence of circumstances which would make such regulation advisable is remote. These four categories (together with "subversive knowledge") form the inner core where regulation should be stoutly resisted.

The second major conclusion proceeding from this classification and analysis is that outside the inner core we have a group of categories ("destructive technology," "slippery slope technology," "economically exploitative technology," "human subjects research," "expensive science," "accidents in science") in which controls are not only conceivable and justified, but where regulations of one sort or another are already in effect. The most important debate within these categories of concerns is not about academic freedom in principle, but about a compromise between the effort to avoid destructive social effects, on the one hand, and the effort to promote scientific and technical creativity and new advances in human welfare, on the other.

The third conclusion which I would draw from this analysis is that even within those categories of concerns where controls are clearly justified, there exists a danger that creativity will be regulated to death by bureaucracies with momentums of their own. Just as we are wary of the "slippery slope" in biomedical ethics, so also should we resist slipping inadvertently into increasing controls over fundamental science, since such controls can easily lead to abuses. Tremendous damage can be caused by the regulators, while the lack of regulation over some types of research would be quite dangerous to society.

In order to achieve a balance between these interests we should look at the categories of concern where controls of some kind seem to be, in principle, both permissible and justified, and then ask the following questions: At what level should controls be imposed? Who should establish and administer the controls? The answers to these questions should be considered separately for each category of research and for each particular problem of regulation which has arisen.

Possible answers to the question, "At what level should controls be imposed?" include the following levels:

1. The individual researcher
2. The laboratory, university, or industry
3. Funding
   a. Government support
   b. Private support
4. Construction of research facilities
5. Actual performance of research
6. Publication of results
7. Application
8. Production
9. Sales
10. Use

Possible answers to the question "Who should establish and administer the controls?" include:

1. The individual researcher
2. The research laboratory, university, or industry
   a. Government
   b. Private
3. A group of researchers in the field
4. An established professional organization
5. Mixed boards of specialists and lay people
6. Established governmental bodies
   a. Local, state, or federal
   b. By means of guidelines or by normal legislation

The main purpose in listing such a large number of levels where regulations could be imposed and on which regulatory agencies might operate is simply to point to the great diversity of possible approaches to individual problems of regulating science and technology; the lists do not represent any sort of "ascending scale" in terms of desirability or undesirability. The levels at which recombinant DNA research has been limited or regulated skipped from "the individual researcher" (before the Asilomar conference) to "a group of researchers in the field" (the Asilomar conference) to "guidelines by federal bodies" (the NIH guidelines). Professional researchers in the field seemed to fear most of all regulation by "local governmental bodies" (such as the Cambridge City Council). If legislation were inevitable, the professional scientists, by and large, preferred a federal law to a local one. Recently, however, they have won increasing support for the view that such legislation is not necessary even on the federal level and that existing guidelines for research are adequate. This issue is still not settled, however, particularly for industrial research laboratories, which are not necessarily affected by the guidelines.

It would be misleading to assume that the professional scientists always prefer regulations or guidelines administered on the federal level rather than local levels, or that a federal approach is always the best one in terms of general societal interests. In the categories "destructive technology" and "accidents in science" local and state regulations play important roles in protecting against fire and in controlling the use and transportation of certain hazardous materials; in the category "slippery slope technology," local advisory committees, often composed of both medical specialists and lay people, have helped to make difficult decisions about the use of life-support machines in hospitals and in recommending priorities in health care.

The question "Who should establish and administer controls?" opens up a whole range of important issues which my analysis has so far not touched. This additional set of questions is essentially political and it involves the problem of the motivation of the critics of science. Some people, upon reading the present essay, would comment that my effort to analyze concerns about science and to evaluate their legitimacy is misdirected, at least in part, because it fails to take into account that many of the critics of science are not really interested in these analytical categories; what they are really interested in is politics. They would argue that such critics of science believe that the present system under which scientists usually control their own work is fundamentally wrong, and that the critics wish to wrest control of science from the scientists and give it to popular nonscientific groups, as the slogan "Science for the People" indicates. Such criticism of my analysis up to this point has some merit, but the political controversy can not be resolved in a useful way, in my opinion, until the categories of concern about science have been viewed separately and until the varying degrees of legitimacy of these concerns have been discerned. Otherwise, the discussion will degenerate into a shouting match in which each side impugns the motives of the other side.

On the question of motivation, it is clear that the knife cuts both ways. If we ask what are the motives of the critics when they attack science, so also can we ask what are the motives of the defenders of science when they reject the critics. The extreme and entirely useless way of approaching the problem is to fall into opposing camps which might be described as "Science for the People" versus "Science for the Scientists." The first group can then easily be written off as political radicals who have no interest in intellectual values; the second group can then be dismissed as elitists who wish to preserve their privileges while having society pay for their work.

The only way out of this dilemma is to recognize that it is an inaccurate portrayal of the existing situation. The greatest value of science is not what it does for scientists, but what it does in both intellectual and material terms for society, of which scientists are one part. It is also clear that society is deeply affected by science and technology in ambiguous and sometimes disturbing ways; some of the public concerns about what might happen if science and technology are left entirely to scientists and engineers are clearly legitimate. In such areas as "slippery slope technology" and "human subjects research" it is obvious that the scientists and engineers directly involved in the work have no monopoly of wisdom about the ethical, psychological, and societal impacts of their work. At the same time, we know that the assertion by lay groups of control over the determination of the inherent value of fundamental research could have disastrous results; history provides some rather graphic examples of how destructive that interference can be.

REFERENCES

[1]I would like to express my appreciation to the Rockefeller Foundation for a Humanities Fellowship for 1976–1977, during which time I did research on the relationship of science and socio-political values, including research for this article. I would also like to thank colleagues at the Program on Science and International Affairs, Harvard University, the members of the MIT seminar on "Limits to Inquiry," and, especially, Harvey Brooks and David Z. Robinson.

[2]Letter of January 18, 1977, from Professor Robert S. Morison and Gerald Holton, quoting Walter Rosenblith, inviting participants to a planning conference for the MIT faculty seminar "Limits to Inquiry."

[3]Opinion Research Corporation, Caravan Surveys, *Attitudes of the U.S. Public Toward Science and Technology: Study II* (Washington, D.C.: National Science Foundation. July, 1974).

[4]Hans Jonas, "Freedom of Scientific Inquiry and the Public Interest," *Hastings Center Report* (August, 1976): 27.

[5]See "An Act for the Development and Control of Atomic Energy," *United States Statutes at Large* (79th Cong., 2nd Sess.), Vol. 60, Washington, D.C., 1946, pp. 755–775. A discussion of the resistance of scientists to the May-Johnson Bill is in Alice Kimball Smith, *A Peril and a Hope: The Scientists' Movement in America: 1945–47* (Chicago: The University of Chicago Press, 1965).

[6]Carl Djerassi, "Probabilities and Practicalities," *Bulletin of Atomic Scientists* (December, 1972): 27.

[7]*Congressional Record* (94th Cong., 1st Sess.) April 9, 1975, VI, pp. H2601–H2602: also, "Bauman Amendment's Chances Down," *Science*, 189 (July 4, 1975): 27.

[8]Robert Sinsheimer, "Comments," *Hastings Center Report* (August, 1976): 18.

[9]Letter to the author from Harvey Brooks (April 12, 1977).

[10]See, for example, W. G. Tifft, W. M. Sinton, and J. B. Priser, "West Ford Dipole Belt: Photometric Observations," *Science*, 141 (August 30, 1963): 798–799.

[11]All quotations in this paragraph are from Cheryl M. Fields, "Who Should Control Recombinant DNA?" *The Chronicle of Higher Education*, XIV, No. 4 (March 21, 1977): 1, 4–5.

[12]"Rassentheorie," *Meyers Neues Lexikon*, VI (Leipzig, 1963), pp. 817–818.

[13]Theodore Roszak, *Where the Wasteland Ends: Politics and Transcendence in Postindustrial Society* Garden City, N.Y.: Doubleday, 1972), pp. *xxiv*, 168.

[14]A. S. Eddington, *Science and the Unseen World* (New York: Macmillan, 1929), p. 49.

[15]Dorothy Nelkin, "The Science-Textbook Controversies," *Scientific American*, 234 (April, 1976): 33–39. Also see her *Science Textbook Controversies* (Cambridge, Mass.: MIT Press, 1977).

[16]"NSF: Congress Takes Hard Look at Behavioral Science Course," *Science*, 188 (May 2, 1977): 426–428.

[17]Cf. Opinion Research Corporation, *op. cit.*, pp. 24–26 for verbatim answers, among which were the following:

People live too fast. Everybody wants too much.
I don't know, but just think so.
Just being directed the wrong way—pollution.
Some manufactured items, such as sprays, etc.
Automobiles have ruined the environment.
Disturbance of the earth. Atomic explosion.
Nobody is an individual. The generation of today's world is becoming more related to a group.
Environment. Pollution. Food pollution. Everything they make is for profit, not for what is best for the people.
The oil spills are harmful. It pollutes the water.
Put people out of work by machines.
Increased the population by cutting down the deaths.
Pollution. Smoking.
Not only are we polluting the earth, we are now polluting space.
People are dying early in life from heart disease caused by stress.
Radiation from the bomb.
The speed factor on automobiles increasing deaths.
I'm disappointed in just about everything they've done.

Spending too much money.
Wasting our money going to the moon.
Going to the moon was wrong.
Open people's eyes—sex education without parents' consent.
Mechanized the whole society.
They change too much too fast—like things in foods and chemicals in stock foods, etc.
Making the LSD drugs that kids get into.
Men go up in space, bringing down some of the diseases.
Generally excludes God.
Don't know.
It has forced a lot of small business people out of work.
Evolution.
Technology created a drug culture.
Invented too many electrical appliances.
I do not believe in prolonging life at death.
They teach children that God didn't make them.
Every time they went to the moon we had bad storms.
Huge cars.

[18]For people who discount the possibility of such a situation, let me portray the following hypothetical scenario. Imagine that you are a scientist in Nazi Germany and that you have just discovered Tay-Sachs disease, an abnormality based on a genetic defect which is more common among Jews than other population groups. Is it not possible that you would try to keep this bit of knowledge away from the eyes of Hitler and National Socialist bureaucrats, and that you might suggest to your trusted colleagues that they do the same?

ROBERT L. SINSHEIMER

The Presumptions of Science

CAN THERE BE "forbidden"—or, as I prefer, "inopportune" knowledge? Could there be knowledge, the possession of which, at a given time and stage of social development, would be inimical to human welfare—and even fatal to the further accumulation of knowledge? Could it be that just as the information latent in the genome of a developing organism must be revealed in an orderly pattern, else disaster ensue, so must our knowledge of the universe be acquired in a measured order, else disaster ensue?

Biological organisms are equipped with many sensors essential to their survival, sensors for heat, cold, pain, thirst, hunger. Social organisms similarly need sensors of peril, particularly as they evolve into new domains—and for these we must use our intelligence, limited as it may be.

Discussion of the possible restraint of inquiry touches a most sensitive nerve in the academic community. If one believes that the highest purpose available to humanity is the acquisition of knowledge (and in particular of scientific knowledge, knowledge of the natural universe), then one will regard any attempt to limit or direct the search for knowledge as deplorable—or worse.

If, however, one believes that there may be other values to be held even higher than the acquisition of knowledge—for instance, general human welfare—and that science and possible other modes of knowledge acquisition should subserve these higher values, then one is willing to (indeed, one must) consider such issues as: the possible restriction of the rate of acquisition of scientific knowledge to an "optimal" level relative to the social context into which it is brought; the selection of certain areas of scientific research as more or less appropriate for that social context; the relative priorities at a given time of the acquisition of scientific knowledge or of other knowledge such as the effectiveness of modes of social integration, or of systems of justice, or of educational patterns.

In short, if one does not regard the acquisition of scientific knowledge as an unquestioned ultimate good, one is willing to consider its disciplined direction. One may, of course, still have grave doubt as to whether mankind can know enough to be able intelligently to guide the rate or direction of the scientific endeavor, but at least one will then accept that we have a responsibility to seek answers—if there be any—to such questions.

The Impact of Science

In 1930 Robert A. Millikan, Nobel Prize winner, founder and long-time leader of Caltech, wrote in an article entitled "The Alleged Sins of Science" that one may "sleep in peace with the consciousness that the Creator has put some foolproof elements into his handiwork, and that man is powerless to do it any titanic physical damage."[1]

To what was Millikan referring? Stimulated by the recombinant DNA controversy, I have looked back to see if there were any similar admonitions or premonitions with respect to the possible consequences of nuclear energy. And there were. Millikan, in 1930, was responding to an earlier writing of Frederick Soddy. In a book entitled *Science and Life* Soddy, who had been a collaborator of Rutherford, had written:

> Let us suppose that it became possible to extract the energy which now oozes out, so to speak, from radioactive material over a period of thousands of millions of years, in as short a time as we pleased. From a pound weight of such substance one could get about as much energy as would be obtained by burning 150 tons of coal. How splendid. Or a pound weight could be made to do the work of 150 tons of dynamite. Ah, there's the rub . . . It is a discovery that conceivably might be made tomorrow in time for its development and perfection, for the use or destruction, let us say, of the next generations, and, which it is pretty certain, will be made by science sooner or later. Surely it will not need this actual demonstration to convince the world that it is doomed if it fools with the achievements of science as it has fooled too long in the past.
>
> War, unless in the meantime man has found a better use for the gifts of science, would not be the lingering agony it is today. Any selected section of the world, or the whole of it if necessary, could be depopulated with a swiftness and dispatch that would leave nothing to be desired.[2]

Millikan commented, just prior to his statement quoted above, "Since Mr. Soddy raised the hobgoblin of dangerous quantities of available subatomic energy [science] has brought to light good evidence that this particular hobgoblin—like most of the hobgoblins that crowd in on the mind of ignorance—was a myth . . . The new evidence born of further scientific study is to the effect that it is highly improbable that there is any appreciable amount of available subatomic energy to tap."[3]

So much for scientific prophecy. But it is indeed instructive and also troubling to recognize that our scientific endeavor truly does rest upon unspoken, even unrecognized, faith—a faith in the resilience, even the benevolence, of nature as we have probed it, dissected it, rearranged its components in novel configurations, bent its forms and diverted its forces to human purpose. Scientific endeavor rests upon the faith that our scientific probing and our technological ventures will not displace some key element of our protective environment and thereby collapse our ecological niche. It is a faith that nature does not set booby traps for unwary species.

Our bold scientific thrusts into *new* territories uncharted by experiment and unencompassed by theory must rely wholly upon our faith in the resilience of nature. In the past that faith has been justified and rewarded, but will it always be so? The faith of one era is not always appropriate to the next, and an unexamined faith is unworthy of science. Ought we step more cautiously as we explore the deeper levels of matter and life?

Most states of nature are quasiequilibria, the outcome of competing forces. Small deviations from equilibrium, the result of natural processes or human intervention, are most often countered by an opposing force and the equilibrium restored, at some rate dependent upon the kinetics of the processes, the sizes of the relevant natural pools of components, and other factors. Although we may therefore speak of the resilience of nature, this restorative capacity is finite and is limited in rate.

For example, if the ozone layer of the atmosphere is lightly and transiently depleted by a nuclear explosion or the atmospheric release of fluorocarbons, the natural processes which generate the ozone layer can restore it to the original level within a brief period. However, should the ozone layer be massively depleted—as by extended, large-scale release of fluorocarbons—many decades would be required for its renewal by natural processes, even if the release of fluorocarbons ceased.

Similiarly, the populations of most living creatures can achieve an equilibrium level dependent upon birth rates and upon death rates from various causes. Most species have an excess capacity for reproduction, so that minor additions to the process of their removal (as by the harvesting of fish) cannot appreciably influence the equilibrium population. Patently however, excessive harvesting removing numbers beyond the reproductive capacity of the species will in time bring about its extinction.

In a similar manner lakes and rivers and air basins can absorb and dispose of limited amounts of pollutant but can be overwhelmed by masses beyond their capacity. Once overwhelmed the very agents responsible for disposal of pollution in small quantities may be destroyed, leaving a "dead" sea.

The concept of resilience extends to the planet as a whole and to the impact upon the manifold equilibria upon which the network of life forms depends as we continue to expand our intensive monoculture agriculture, as we continue to increase the total of human energy consumption (the man-made release of energy in the Los Angeles basin is now estimated at about 5 percent of the solar input), as we continue to raise the atmospheric level of CO_2 by combustion of fossil fuel, and so forth.

Because human beings (and most creatures) are adapted by evolution to the near equilibrium states, the resilience provided by the restorative forces of nature has appeared to us to be not only benevolent, but unalterable. Less overt than our faith in the resilience of nature is the faith with which we have relied upon the resilience of our social institutions and their capacity to contain the stress of change and to adapt the knowledge gained by science—and the power inherent in that knowledge—to the benefit of society, more than to its detriment. The fragility of the equilibria underlying social institutions is even more apparent than of the equilibria of nature. Political, economic, and cultural balances have shifted drastically in human history under the impact of new technologies, or new ideologies or religions, of invading peoples, of resource exhaustion, and other changes. Our faith in the resilience of both natural and man-made phenomena is increasingly strained by the acceleration of technical change and the magnitude of the powers deployed.

Physics and chemistry have given us the power to reshape the physical nature of the planet. We wield forces comparable to, even greater than those of, natural catastrophes. And now biology is bringing to us a comparable power over the world of life. The recombinant DNA technology, while significant and potentially a grievous hazard in itself (through the conceivable production, by design or by inadvertance, of new human, animal, or plant pathogens or of novel forms capable of disrupting important biological equilibria), must be seen as a portent of things to come.

The present recombinant DNA technology, which permits the addition or replacement of a few genes in living cells, is but the first prototype of genetic engineering. More powerful means involving cell fusion or chromosome transfer are already close to hand; even more sophisticated future developments appear assured. Since genes determine the basic structures and biological potentials of all living forms, the ultimate potential of genetic engineering for the modification and redesign of plants and animals to meet human needs and desires seems virtually unlimited.

Such capabilities will pose major questions as to the extent to which mankind will want to assume the responsibility for the life forms of the planet. Further, there is no reason to believe the same technology will not be applicable to mankind as well; the capability of human genetic engineering will raise profound questions of values and judgment for human societies.

It seems paradoxical that a living organism emergent from the evolutionary process after billions of years of blind circumstance should undertake to determine its own future evolution. The process is perhaps analogous to that of the mind seeking to understand itself. In both cases it is uncertain whether the attempt can possibly be successful. Nonetheless, at this point perhaps we had best step back and reconsider what it is we are about.

For four centuries science has progressively expanded our knowledge and reshaped our perception of the world. In that same time technology has correspondingly reshaped the pattern of our lives and the world in which we live.

Most people would agree that the net consequence of these activities has been benign. But it may be that the conditions which fostered such a benign outcome of scientific advance and technological innovation are changing to a less favorable set. Changes in the nature of science or technology or in the external society—in either the scale of events or their temporal order—can affect the preconditions, the presumptions, of scientific activity, and can thus alter the future consequences of such activities.

Both quantitative and qualitative changes have surely affected the impact of science and technology upon society. Quantitatively, the exponential growth of scientific activity and the unprecedented magnitude of modern industrial ventures permit the introduction of new technologies (e.g., fluorocarbon sprays) on a massive scale within very brief periods often with unforeseen consequence. Qualitatively, science and technology have been directed increasingly to synthesis—to the formulation of new substances designed for specific human purpose. Thus we have synthetic atoms (plutonium, strontium-90), synthetic molecules (dioxin, kepone, DDT) and now synthetic microorganisms (recombinant DNA). In these activities we introduce wholly novel substances into the planetary environment, substances with which our evolution has not always prepared us to cope.

Can we continue to rely upon the past four centuries as a guide for scientific activity, given these changes? Other human activities of this same era are now increasingly seen in a different hue. The same period witnessed exponential increases in population and in the exploitation of natural resources for material wealth. Few would argue continuance of such trends will be benign.[4] The same era has witnessed the constant acceleration of the rate of change, the increasing dominance of technology in the affairs of men.

The constantly accelerating accretion of knowledge, therefore, may not always be counted as a good. Can circumstances change so as to devalue the net worth of new knowledge? Might a pause or slowdown for consolidation and reflection then be more in order? Indeed, could it be that some knowledge could, at this time, be positively malign? Hard questions, perhaps not answerable, perhaps not the right questions, but they are not answered for 1977 by invoking Galileo or Darwin or Freud. I believe they demand our thought.

I would advance for consideration some propositions that frankly I'm not at all sure I entirely believe. I think that in order to find out what one does believe it is necessary to go beyond what one can readily accept—to explore honestly more extreme and more remote positions so that one's position is based upon intelligent choice, not simple ignorance.

The domain I propose to explore can be indicated by a question. The question is one I have actually raised within the administration at Caltech (and it could as well be raised elsewhere). Institutions such as Caltech and others devote much energy and effort and talent to the advancement of science. We raise funds, we provide laboratories, we train students, and so on. In so doing we apply essentially only one criterion—that it be good science as science—that the work be imaginative, skillfully done, in the forefront of the field. Is that, as we approach the end of the twentieth century, enough? As social institutions, do Caltech and others have an obligation to be concerned about the likely *consequences* of the research they foster? And if so, how might they implement such a responsibility?

For reasons which probably need no elaboration Caltech has been more than reluctant to come to grips with this question. And, indeed, it just may be—and I say this with real sorrow—that scientists are simply not the people qualified to cope with such a question. The basic tactic of natural science is analysis: fragment a phenomenon into its components, analyze each part and process in isolation, and thereby derive an understanding of the subject. In physics, chemistry, even biology, this tactic has worked splendidly.

To answer my question, however, the focus must not be inward but outward, not narrowed but broadened. The focus must be on all the ties of the sciences to society and culture and on the impact of scientific knowledge and technological advancement on all human, indeed all planetary, life.

Consider as an instance the recombinant DNA issue. The natural tendency of the scientist, if he will admit this a problem, is to break it down, to decompose it into individually analyzable situations. If there is a danger, quantitate it: what is the numerical chance of the organisms escaping, of their colonizing the gut, of their penetrating the intestinal epithelium, of their causing disease (what disease)? If you point out that there is a nearly infinite set of possible scenarios of misfortune—that accidents do happen and in unpredictable ways, that humans do err, that bacterial or viral cultures do become contaminated, that indeed aspects of this technology involve inherently unpredictable consequence and hence are not susceptible to quantitative analysis—you are regarded as unscientific.

The consequences of the interaction of known but foreign gene products with the complex contents of a bacterial cell would be difficult enough to predict, much less the consequences of the interactions of unknown gene products, as produced in "shotgun" experiments. Some of these consequences may well modify, in unpredictable ways, the likelihood of the organism's survival or persistence in various environments, its potential toxicity for a host or nearby life forms. It may alter, for instance, an organism's survival in an animal intestine, contrary to our expectations, for we have presumed that we know all factors important for survival there and that no new successful adaptations could emerge.

For complex reasons, consideration of the potential hazards from organisms with recombinant DNA has focused upon immediate medical concerns. That these organisms with unpredictable properties might have impact upon any of the numerous microbiological processes which are important components of our life support systems is simply dismissed as improbable. The fact that these organisms are evolutionary innovations and have within themselves, as do all living forms, the capacity (if they survive) for their own unpredictable future evolutionary development is ignored, or dismissed as mystical.

If you point out that the recombinant DNA issue simply cannot be effectively considered in isolation but must be viewed in perspective and in a larger context as a possible precursor to future technologies available to many elements of society (including totalitarian governments, the military, and terrorist factions) your remarks are regarded as irrelevant to science.

There is an intensity of focus in the scientific perspective which is both its immediate strength and its ultimate weakness. The scientific approach focuses rigorously upon the problem at hand, ignoring as irrelevant the antecedents of motive and the prospectives of consequence.

Viewed objectively such an approach can only make sense if either (1) the consequences are always trivial, which is patently untrue, or (2) the consequences are always benign, that is, if the acquisition of knowledge, of any knowledge at any time, is always good, a proposition one might find hard to defend, or (3) the dangers and difficulties inherent in any attempt to restrict the acquisition of knowledge are so great as to make the unhindered pursuit of science the lesser evil.

In thinking about the impacts of science, we should, perhaps, reflect upon the inverse of the uncertainty principle. Perhaps it might be called the certainty principle. The uncertainty principle is concerned with the inevitable impact of the observer upon the observed, which thereby alters the observed. Conversely, there is an effect of the observed upon the observer. The discovery of new knowledge, the addition of new certainty, which correspondingly diminishes the domain of uncertainty and mystery, inevitably alters the perspective of the observer. We do not see the world with the same eyes as a Newton or a Descartes, or even a Faraday or a Rutherford.

The acquisition of a discipline sharpens our vision in its domain, but too frequently it seems also to blind us to other concerns. Thus immersion in the world of science, with its store of accumulated and substantiated fact, can make the participant intolerant of, and impatient with the uncertainties and nonreproducibilities of the human world. Engrossed in the search for knowledge, scientists tend to adopt the position that more knowledge is the key to the solution to human problems. They may not see that the uses we make of knowledge or the ways in which we organize to use knowledge can, as well, be the limiting factors to the human condition, and they forget that even within science our knowledge and our theories are always human constructs. Moreover, we should always remember (lest we become too secure and even smug) that our knowledge and our theories are ever incomplete.

Of Dubious Merit

To make this discussion more specific let me consider three examples of research that I personally consider to be, on balance, of dubious merit. One is in an area of rather applied research, the second in a very speculative but surely basic area, and the third in the domain of biomedical research, which we most often conceive to be wholly benign.

The first I would cite is current research upon improved means for isotope fractionation. In one technique, one attempts to use sophisticated lasers[5] to activate selectively one isotope of a set. I do not wish to discuss the technology but rather the likely consequence of its success. To be sure, there are benign experiments that would be facilitated by the availability of less expensive, pure isotopes. For some years I wanted to do an experiment with oxygen 18 but was always deterred by the cost.

But does anyone doubt that the most immediate application of isotope fractionation techniques would be the separation of uranium isotopes? This country has recently chosen to defer, at least, if not in fact to abandon, the plutonium economy and the breeder reactor because of well-founded concern that plutonium would inevitably find its way into weapons. We are thus left with uranium-fueled reactors. But uranium 235 can also be made into a bomb. Its use for power is safer only because of the difficulty in the separation of uranium 235 from the more abundant uranium 238. If we supersede the complex technology of Oak Ridge, if we devise quick and ingenious means for isotope separation, then one of the last defenses against nuclear terror will be breached. Is the advantage worth the price?

A second instance I would cite of research of dubious merit, and one probably even more tendentious than the first, relates to the proposal to search for and contact extraterrestrial intelligence.[6] Recent proposals suggest that, using advanced electronic and computer technology, we could monitor a million "channels" in a likely region of the electromagnetic spectrum, "listening" over several years for signals with an "unnatural" regularity or complexity.

I am concerned about the psychological impact upon humanity of such contact. We have had the technical capacity to search for such postulated intelligence for less than two decades, an instant in cosmic terms. If such intelligent societies exist and if we can "hear" them, we are almost certain to be technologically less advanced and thus distinctly inferior in our development to theirs. What would be the impact of such knowledge upon human values?

Copernicus was a deep cultural shock to man. The universe did not revolve about us. But God works in mysterious ways and we could still be at the center of importance in His universe. Darwin was a deep cultural shock to man. But we were still number one. If we are closer to the animals than we thought before, and through them to the rocks and the sea, it does not really devalue man to revalue matter. To really be number two, or number 37, or in truth to be wholly outclassed, an inferior species, inferior on our own turf of intellect and creativity and imagination, would, I think, be very hard for humanity.

The impact of more advanced cultures upon less advanced has almost invariably been disastrous to the latter. We are well acquainted with such impacts as the Spanish upon the Aztecs and Incas or the British and French upon the Polynesians and Hawaiians. These instances were, however, compounded by physical interventions (warfare) and the introduction of novel diseases. I want to emphasize the purely cultural shock. Hard learned skills determinant of social usefulness and positions become quickly obsolete. Less advanced cultures quickly become derivative, seeking technological handouts. What would happen to *our* essential tradition of self-reliance? Would we be reduced to seekers of cosmic handouts?

The distance of the contacted society might, to some degree, mitigate its consequent impact. A contact with a round trip communication time of ten years would have much more effect than one with a thousand years. The likelihood of either is, however, a priori, unknown. Nor is it inconceivable that an advanced society could devise means for communication faster than light.

The proponents of such interactions have considered the consequences briefly. In a 427-page book *Communication with Extraterrestrial Intelligence*[7] sixteen pages comprise a chapter entitled "Consequences of Contact." Opinion therein ranges from "Our obligation is, I feel, to stress that in any sensible way this problem has no danger for human society. I believe we can give a full guarantee of this" to "If we come in contact with some superior civilization this would mean the end of our civilization, although that might take a while. Our period of culture would be finished."

How and by whom should such a momentous decision[8] be made—one that will clearly, if successful, have an impact upon all humanity? Somehow I cannot believe it should be left to a small group of enthusiastic radioastronomers.

My concern here does not extend so far that I would abolish the science of astronomy. If the astronomers in the course of their science come across phenomena that can only be understood as the product of intelligent activity, so be it. But I do not believe that is the same as deliberately setting out to look for such activity with overt pretensions of social benefit.

The third example of research I consider of dubious merit concerns the aging process. I would suggest this subject exemplifies in supreme degree the eternal conflict between the welfare of the individual and the welfare of society and, indeed, the species. Obviously, as individuals, we would prefer youth and continued life. Equally obviously, on a finite planet, extended individual life must restrict the production of new individuals and that renewal which provides the vitality of our species.

The logic is inexorable. In a finite world the end of death means the end of birth. Who will be the last born?

If we propose such research we must take seriously the possibility of its success. The impact of a major extension of the human life span upon our entire social order, upon the life styles, mores, and adaptations associated with "three score and ten," upon the carrying capacity of a planet already facing overpopulation would be devastating. At this time we hardly need such enormous additional problems. Research on aging seems to me to exemplify the wrong research on the wrong problem in the wrong era. We need that talent elsewhere.

Is Restraint Feasible?

If one concedes, however reluctantly, that restraint of some directions of scientific inquiry is desirable, it is appropriate to ask if it is feasible and, if so, at what cost.

Some of my colleagues, not only in biology but in other fields of science as well, have indicated to me that they too increasingly sense that our curiosity, our exploration of nature, may unwittingly lead us into an irretrievable disaster. But they argue we have no alternative.[9] Such a position is, of course, a self-fulfilling prophecy.

I would differentiate among what might be called physical feasibility, logical feasibility, and political feasibility.

I believe that actual physical restraint is in principle feasible. There are two evident avenues of control: the power of the purse and access to instruments. Control of funding is indeed already a powerful means for control of the directions of inquiry for better or worse. To the extent that there exists a multiplicity of sources of support, such control is porous and incomplete, but it is clearly a first line of restraint.

Research today cannot be done with household tools. It is difficult to imagine, for instance, any serious research on aging that would not require the use of radioisotopes or an ultracentrifuge or an electron microscope. The use of isotopes is already regulated for other reasons. Access to electron microscopes could, in principle, be regulated, albeit at very real cost to our current concepts of intellectual freedom.

An immediately related, important aspect of any policy of restraint concerns the distinctions to be made about the nature of research. Can we logically differentiate research on aging from general basic biologic studies? I expect we cannot in any simple, absolute sense. Yet obviously the people who established the National Institute of Aging must have believed that there is a class of studies which deserves specific support under that rubric. Indeed, distinctions of this sort are made all the time by the various institutes of National Institutes of Health in deciding which grant applications are potentially eligible for their particular support. Pragmatically, and with some considerable margin of error, such distinctions can be and are made.

It is frequently claimed that the "unpredictability" of the outcome of research makes its restraint, for social or other purpose, illogical and indeed futile. However, the unpredictability of a research outcome is not an absolute but is both quantitatively and qualitatively variable.

In more applied research within a field with well-defined general principles, the range of possible outcomes is surely circumscribed. In more fundamental research, in wholly new fields remote from prior human experience—as in the cosmos, or the subatomic world, or the core of the planet—wholly novel phenomena may be discovered. But, for instance, even in a fundamental science such as biology, most of the overt phenomena of life have been long known.

The basic principles of heredity were discovered by Mendel a century ago and were elaborated by Morgan and others early in this century. The understanding of genetic mechanism, the reduction of genetics to chemistry, had to await the advent of molecular biology. This understanding of mechanism has now provided the potential for human interventions, for genetic engineering, but it has not significantly modified our comprehension of the genetic basis of biological process.[10]

The path of modern biology will surely lead to further understanding of biological mechanism, with subsequent application to medicine and agriculture (and accompanying social impact). But it would seem likely that only within the central nervous system may there be the potential for wholly novel—and correspondingly wholly unpredictable—process. Even there, the facts of human psychology and the subjective realities of human consciousness have long been familiar to us, albeit the underlying mechanisms are indeed obscure.

Political feasibility is, of course, another question. The constituency most immediately affected is, of course, the scientific. And despite our protestations and alarms this community does have real political influence. It would seem unlikely to me that a policy of scientific restraint could be adopted in any sector unless a major portion of the scientific community came to believe it desirable.

For this to happen, that community will clearly have to become far more alert to, and aware of, and responsible for the consequences of their activities. The best discipline is self-discipline. Scientists are keenly sensitive to the evaluations of their peers. The scientific community and the leaders of our scientific and technical institutions will have to develop a collective conscience; they will have to let it be known certain types of research are looked upon askance, much as biological warfare research is today; it needs to be understood that such research will not be weighed in considerations of tenure and promotion; societies need to agree not to sponsor symposia on such topics. All of these and similar measures short of law could indeed be very effective.

I am well aware of the dangers implicit in such forms of cultural restraint. But I think we really must look at the dangers we face in the absence of self-restraint. Do we accept only the restraint of catastrophe?

If we are to consider this position, we must do so in a forthright manner. We must be willing to explore the vistas exposed if we lower conventional taboos and sanctions. We may not at first enjoy what we see, but at least we will have a better perception of the available alternatives. Any attempt to limit the freedom of scientific inquiry will surely involve what will appear, at least at first, to be quite arbitrary distinctions—judgmental decisions, the establishment of boundaries in gray and amorphous terrain. These are, however, familiar processes in our society, in the courts, in the legislatures. Indeed, most of us are familiar with such problems in our educational activities. The selection of new faculty, the award of tenure, the assignment of grades are clearly judgmental decisions.

In science we try with some success to elude the necessity for such very human judgments. Indeed, one suspects that many persons go into science precisely to avoid the necessity for such complex decisions—in search of a domain of unique and unequivocal answers of enduring validity. And it is painful to see the sanctuary invaded.

Admittedly it is difficult to achieve consensus on the criteria for judgmental decisions. Such consensus is all the more difficult in the sphere of international activities such as science which involve participants from diverse cultures and traditions.

Conversely there are many persons who prefer the more common, perhaps the more human world of ambiguity and compromise and temporally valid judgments and who resist the seemingly brutal, life and death, cataclysmic types of decision increasingly imposed upon society by the works of science. And science and scientists cannot stand wholly aloof from these latter dilemmas—for science is a human activity and scientists live in the human society. We cannot expect the adaptation to be wholly one-sided.

Even if, at best, we can only slow the rate of acquisition of certain areas of knowledge, such a tactic would give us more time to prepare for social adaptation—if we mobilize ourselves to use that time.

The Case for Restraint

The view one exposes by lifting that sanction we label freedom of inquiry is frankly gloomy. It would seem that we are asked to make thorny decisions and delicate differentiations, to relinquish long-cherished rights of free inquiry, to forego clear prospects of technological progress. And it would seem that all these concessions stem ultimately from recognition of human frailty and from recognition of the limitations of human rationality and foresight, of human adaptability and even good will. Just such recognitions have already spawned many of our institutions and professions—religions, the law, government, United Nations—yet all of these are as imperfect as the world they are designed to restrain and improve.

At each level of human activity, whether individual, group, or national, we continually struggle to find acceptable compromises between the freedom to pursue varied courses and goals and the conflicts that arise when one person's actions run contrary to another's. In a crude sense the greater the power available to an entity, the more limitations must be imposed upon its freedom if conflict is to be averted. Ideally such limits are internalized through education and conscience, but we all understand the inadequacy of that process.

In short, we must pay a price for freedom, for the toleration of diversity, even eccentricity. That price may require that we forego certain technologies, even certain lines of inquiry where the likely application is incompatible with the maintenance of other freedoms. If this is so and if we can recognize and understand this, perhaps we can, as scientists, be more accepting.

Some will argue that knowledge simply provides us with more options and thus that the decision point should not be at the acquisition of knowledge but at its application.

Such a view, however ideal, overlooks the difficulty inherent in the restriction of application of new knowledge, once that knowledge has become available in a free society. Does anyone really believe, for instance, that knowledge permitting an extension of the human life span would not be applied once it were available?

One must also recognize again that the very acquisition of knowledge can change both the perceptions and the values of the acquirer. Could, for instance, deeper knowledge of the realities of human genetics affect our commitment to democracy?

It may be argued that the cost, however it may be measured, of impeding research would be greater to a society than the cost of impeding application. Perhaps so. This issue could be debated, but it must be debated in realistic terms with regard for the nature of real people and real society and with full understanding that knowledge is indeed power.

Although the nature of the measures necessary to restrict the application of knowledge has seldom been analyzed, the measures needed would surely be dependent upon the size of investment required to apply the knowledge, as well as on the form of and the need for the potential benefits of the knowledge, among other things. The compatibility of such restrictive measures with the principles of a democratic society would need to be considered. Restriction of nuclear power may be a case in point.

Alvin Weinberg has developed the concept of the technological fix as the simple solution to cut the Gordian knot of complex social problems. However, we seem to be discovering that the application of one technological fix seems to lead us into another technological fix. For example, the development of antibiotics and other triumphs of modern medicine has led to the tyranny of overpopulation. In efforts to cope with overpopulation by more intensive agriculture, we develop pesticides, herbicides and other chemicals which increase the level of environmental carcinogenesis. And so on.

The moral is that we cannot ignore the social and cultural context within which the technology is deployed. In retrospect we can see that in the cultural and social context of the seventeenth, eighteenth, and nineteenth centuries the consequences of technological innovation were most often benign. Whether because of change in the society and culture or change in the nature and effectiveness of technology, at some time in the twentieth century the balance began to shift and by now our addiction to technology begins to assume an unpleasant cast.

We are indeed addicted to technology. We rely ever more upon it and thus become its servant as well as its master. It has led to human populations insupportable without its aid. Further, new technologies shape our perceptions; they spawn expectations of change or stir deep fears of disaster. They dissociate us from the past and becloud the shape of the future. Even the oldest boundary conditions of humanity fall as we leave the planet and as we plan to reshape our genes.

Our academic institutions and our professional societies foster and promote science. To some degree they also have concern for its consequences, but it is a minor aspect. The principle that one should separate agencies which promote and agencies which regulate may apply here.

But where then is the balance, the necessary check to the force of scientific progress? Is the accumulation of knowledge unique among human activities—

an unmitigated good that needs no counterweight? Perhaps that was true when science was young and impotent, but hardly now. Yet we lack the institutional mechanisms for regulation.

Our experience with constraint upon science has hardly been encouraging. From the Inquisition to Lysenko such constraint has been the work of bigots and charlatans. Obviously, if it is to be done to a good purpose, any restraint must be informed, both as to science and as to the larger society on which science impacts.

The acquisition of knowledge is a human, a social, enterprise. If we, through the relentless, single-minded pursuit of new knowledge so destabilize society as to render it incapable—or unwilling—to continue to support the scientific enterprise, then we will have, through our obsession, defeated ourselves.

At Caltech and the many other academic institutions, we have now, *culturally*, cloned Galileo a millionfold. We have nurtured this Galilean clone well; we award prizes and honors to those most like the original. No doubt this clone has been most beneficial for humanity, but perhaps there is a time for Galileos. Perhaps we need in this time to start another clone.

REFERENCES

[1]R. A. Millikan, "Alleged Sins of Science," *Scribners Magazine*, 87 (2) (1930): 119–130.

[2]Frederick Soddy, *Science and Life* (London: John Murray, 1920).

[3]Precisely what evidence Dr. Millikan had in mind is uncertain. However, it was generally appreciated that the efficiency of nuclear transformation by the charged particles then in use was so low that there was no significant prospect of a net release of energy. No practical chain reaction could yet be envisaged.

[4]A. V. Hill in his presidential address to the British Association for the Advancement of Science in 1952, referring to the population problem, said, "If ethical principles deny our right to do evil in order that good may come, are we justified in doing good when the foreseeable consequence is evil?"

[5]See A. S. Krass, "Laser Enrichment of Uranium: The Proliferation Connection," *Science*, 196 (1977): 721–731; also B. M. Casper, "Laser Enrichment: A New Path to Proliferation?" *Bulletin of Atomic Scientists*, 33 (1) (1977): 28–41.

[6]See T. B. H. Kuiper and M. Morris, "Searching for Extraterrestrial Civilizations," *Science*, 196 (1977): 616–621; also B. Murray, S. Gulkis, and R. E. Edelson, "Extraterrestrial Intelligence: An Observational Approach," *Science*, in press.

[7]C. Sagan (ed.), *Communication with Extraterrestrial Intelligence (EETC)* (Cambridge, Mass.: MIT Press, 1973).

[8]Conceivably, we might not be given this choice if an advanced civilization were determined to contact us. At present however, it would seem to be our option.

[9]This is not a new perception. "The world is now faced with a self-evolving system which it cannot stop. There are dangers and advantages in this situation. . . . Modern science has imposed upon humanity the necessity for wandering. Its progressive thought and its progressive technology make the transition through time, from generation to generation, a true migration into uncharted seas of adventure. The very benefit of wandering is that it is dangerous and needs skill to avert evils. We must expect, therefore, that the future will disclose dangers. It is the business of the future to be dangerous; and it is the merit of science that it equips the future for its duties," wrote A. W. Whitehead in *Science and the Modern World*.

[10]Indeed the failure to discover a new class of phenomena underlying genetics has been most disappointing to some. See Gunther S. Stent, "That Was the Molecular Biology That Was," *Science*, 160 (1968): 390–395.

DAVID BALTIMORE

Limiting Science: A Biologist's Perspective

CONTEMPORARY RESEARCH in molecular biology has grown up in an era of almost complete permissiveness. Its practitioners have been allowed to decide their own priorities and have met with virtually no restraints on the types of work they can do. Viewed as a whole, the field has not even met with fiscal restraints because, relative to "big science," molecular biology has been a relatively cheap enterprise.

Some of the funds that fueled the initial, seminal investigations in molecular biology were granted because of the medical implications of work in basic biology. Most continue to come from agencies concerned with health. Although basic research in biology has yet to have a major impact on the prevention and treatment of human disease, a backlash already seems to be developing in which various groups in our society question whether the freedom that has characteristically been granted to research biologists by a permissive public requires modification. Among the numerous elements prompting this questioning are impatience with the lack of practical results, and fears that direct hazards might result from the experimentation, that basic research may not be an appropriate investment of significant funds, and that dangerous new technologies may flow from discoveries in basic biology. Lay critics as well as a few members of the profession have argued that molecular biologists should concentrate their efforts in certain areas of research, like fertility mechanisms, while other areas, like genetics or aging, are possibly dangerous and certainly not worthy of financial support from the public. These critics believe that they can channel contemporary biology to fit their own conception of appropriate research.

I wish to argue that the traditional pact between society and its scientists in which the scientist is given the responsibility for determining the direction of his work is a necessary relationship if basic science is to be an effective endeavor. This does not mean that society is at the mercy of science, but rather that society, while it must determine the pace of basic scientific innovation, should not attempt to prescribe its directions.

What we call molecular biology today had its origins in individual decisions of a small number of scientists during the period from the late 1930s through the early 1950s. These people were trained in diverse fields, among them physics, medicine, microbiology, and crystallography. They created molecular biology out of the realization that the problems posed by genetics were central to understanding the structures of living systems. No one channeled them towards this line of thinking, no one cajoled them to tackle these problems; rather, their own curiosity and sense of timing led them to try to elucidate these mysteries. This history provides a model of how the most effective science is done.

It is partly the successes of molecular biology that have brought on the questioning of whether scientists should be allowed their freedom of decision. It is therefore worthwhile tracing the development of concerns about whether certain areas of science should be closed to investigation. Molecular biology is the science that has revealed to us the nature of one of the most fundamental of all substances of life, the gene. There is a very simple underlying reality to the transmission of characteristics from parents to children: a code based on four different chemicals, denoted by the letters A, T, G, and C, is used to store the information of heredity. The order in which these four chemicals appear in a virtually endless polymer, called DNA, is the language of life.

Knowing that DNA was the physical storehouse of the genes, Watson and Crick in 1953 first solved the problem of how the DNA is organized to assure that information is transferred with almost perfect reliability from parent cells to progeny cells. They showed that DNA is made of two strands, intertwined together into a helix, and also that the four subunits in DNA are physically related so that only A and T form a specific pair as do G and C. The two strands are held together by this specific pairing so that the two strands carry redundant information. Each gene is thus really a gene and a mirror-gene; this self-complementarity allows DNA to repair itself—and therefore maintain the fidelity of its information—and also to duplicate itself, sending identical copies into the two daughter cells resulting from a round of cell division.

Following the monumental discovery of the structure of DNA, many scientists have contributed to learning how to make DNA, how to read DNA, how to cut DNA, how to rejoin DNA, and in general, how to manipulate at will the genes of very simple organisms.

What good has all of this new knowledge brought to the average person? First, and of great importance, is the contribution that scientific advances make to human culture. The continued accumulation of knowledge about ourselves and our environment is a crucial cultural aspect of contemporary life. Science as well as art illuminates man's view of himself and his relation to others. Our knowledge of how we work, how one person differs from and is similar to another, what is health and what disease, what we need to support health, etc., helps to set the ground rules for the debates of politics and the productions of art.

The more practical benefits of biology can be expected to come in the future in the form of medical advances, or increases in food production, or in other manipulations of life processes that will be able to provide positive contributions to civilization. But molecular biology, for all of its power as a basic science, has not been easily translated into tangible benefits. This is a situation that could change very soon. New discoveries are rapidly bringing molecular biology closer to an ability to affect the lives of the general public.

Recombinant DNA

Of the advances that have occurred, a critical one has been the development of a process called recombinant DNA research. This is a technique whereby different pieces of DNA are sewn together using enzymes; the chimeric DNA is then inserted into a bacterium where it can be multiplied indefinitely. Because the method allows genes from any species in the world to be put into a common type of bacterium, there is a theoretical possibility of hazard in this research. The potential for unforeseen occurrences led a number of scientists, including me, to issue in 1974 a call for restraint in the application of these new methods. We were addressing a limited problem, whether there could be a recognizable hazard in the performance of certain experiments. That limited question opened a floodgate; other questions came pouring out and are still coming. They have led to front cover stories in weekly magazines, to serious attempts at federal legislation, to a demoralization of some of the community of basic research biologists, and, most significantly, to a deep questioning of whether further advances in biology are likely to be beneficial or harmful to our society.

Much of the discussion about recombinant DNA research has centered on whether the work is likely to create hazardous organisms. The mayor of Cambridge, Massachusetts, raised the specter of Frankenstein monsters emerging from MIT and Harvard laboratories, and speculations about the possibility of inadvertent development of a destructive organism like the fictitious Andromeda Strain have been much in the news. I am personally satisfied that most of such talk is simply science fiction and that the research can be made as safe as any other research. The people who understand infectious diseases best make the arguments most strongly that recombinant DNA research is not going to create monsters. But rather than defend my judgment that the safety issue has been blown out of proportion, I want to consider some of the more general issues that have been raised by the controversy.

Genetic Engineering

If safety were the most important consideration behind the debate about recombinant DNA then we might expect the debate to focus on the hazards of doing recombinant DNA experiments. Instead, many of the discussions that start considering such questions, soon turn to a consideration of genetic engineering.

Two techniques labeled genetic engineering exist. Both originated because not everybody's genes are perfectly designed for the job of being a functioning human being, so that many instances of blood disorders, mental problems, and a host of other disabilities are traceable to a malfunctioning gene. It would be a triumph of medicine if the effects of such genes could be countered, and two approaches for countering them have been considered, both of which are called *genetic engineering*. One approach involves altering some cells of the body so that they can carry out the needed function. A patient could, for instance, be treated in this way for a blood disease caused by an abnormal protein made by a mutant gene. A normal gene would be inserted into the precursor cells—immature bone marrow cells that ultimately develop into functioning blood cells. In this way, a normal protein could be made in place of, or along with, the aberrant protein. The genetically altered blood cell precursor could then cure the patient's disease. But the malfunctioning gene would still be transmitted to the patient's offspring. Because this form of genetic engineering would not change the gene pool of the species and because it may prove an effective medical treatment of disease, it does not present the same moral problem as the other form. It is likely to be the first type of genetic engineering tried on human beings, and might be tried within the next five years.

The second type of genetic engineering presents more of a dilemma because it could change the human gene pool. This would involve replacement of genes in the germ cells, cells that transmit their genes to our offspring. Such a change would represent a permanent alteration of the types of the genetic information that constitutes our species. Replacement of germ cell genes would be very difficult and is, I suspect, at least twenty years away. It presents no theoretical problems, only formidable logistic problems.

Both forms of genetic engineering, but especially the engineering of germ cells, present two very deep and perplexing problems: who is to decide, and how shall they decide what genes are malfunctional? Decisions about which genes are good and which bad truly represent decisions of morality and are therefore highly subjective. Fear that dictators will decide which genes should be suppressed and which promoted, and that their criteria for decision will be how best to maintain their own power, has made the phrase "genetic engineering" symbolic of the moral problems that can be created by modern biology.

Genetic engineering of human beings is not the same as recombinant DNA research. Genetic engineering is a process carried out on human beings; recombinant DNA methods are ways to purify genes. Such genes might be used in genetic engineering procedures but are much more likely to be used as tools in studies of biological organization or as elements in a biological manufacturing process. Although genetic engineering is not the same as recombinant DNA research, the two are rightly linked, because recombinant DNA work is hastening the day when genetic engineering will be a feasible process for use on certain human diseases. Since recombinant DNA work is also bringing closer the discovery of many other possible new medical treatments, and is likely to bring other new capabilities, why is there so much focus on genetic engineering? I believe that genetic engineering, because of its tabloid appeal, has become a symbol to many people of the frightening potential of modern-day technology. Rather than seeing in molecular biology the same complex mixtures of appropriate and inappropriate applications that characterize all powerful sciences, many people have allowed the single negative catch phrase, genetic engineering, to dominate discussions. People worry that if the possibility of curing a genetic defect by gene therapy should ever become a reality, the inevitable result would be "people made to order." It is argued that unless we block recombinant DNA research now we will never have another chance to control our fate.

Limits to Basic Biology

To see the form of the argument against recombinant DNA research most clearly and to highlight its danger to intellectual freedom and creativity, we should realize that similar arguments have been put forward relative to other areas of basic biology. One of the most respected critics of recombinant DNA research, Robert Sinsheimer, has for instance, made comparable analyses of two other research topics, research on aging and attempts to contact extraterrestrial beings.[1] He argues that if research on aging were to be successful, people could live to much older ages and the changed age structure of the population would bring serious stresses to society. His fear about searching for extraterrestrial beings is, again, that we could be successful, and he considers that the discovery of civilizations much more advanced than ours would have effects on us like those the Europeans had on native Americans after the discovery of the New World. In the case of recombinant DNA research as well as the other topics, Dr. Sinsheimer says we should avoid studying these areas of science. Rather, he believes, we should put our resources into investigating areas of proven need, such as fertility control.

These examples—and I could choose many others, especially outside of biology— have the same general form: certain people believe that there are areas of research that should be taboo because their outcome might be, or in some scenarios will be, detrimental to the stable relationships that characterize contemporary society. I have heard the argument in a different and more pernicious form from members of a Boston area group called Science for the People. They argue that some research in genetics should not go on because its findings might be detrimental to the relationships they believe should characterize a just society. Such arguments are reminiscent of those surrounding the eugenics movements that developed in Germany and Russia in the 1920s. After a period of intense debate, these countries with opposing ideologies settled on opposing analyses of the role of genetics in determining human diversity.[2] German scientists and politicians espoused a theory of racial purification by selective breeding, while Russia accepted the Lamarckian principle of transmission of acquired characteristics. In both cases, science was forced into a mold created by political and social ideology, and in both cases the results were disastrous.

Necessity of Freedom

As I see it, we are being faced today with the following question: should limits be placed on biological research because of the danger that new knowledge can present to the established or desired order of our society? Having thus posed the question, I believe that there are two simple, and almost universally applicable, answers. First, the criteria determining what areas to restrain inevitably express certain sociopolitical attitudes that reflect a dominant ideology. Such criteria cannot be allowed to guide scientific choices. Second, attempts to restrain directions of scientific inquiry are more likely to be generally disruptive of science than to provide the desired specific restraints. These answers to the question of whether limits should be imposed can be stated in two arguments. One is that science should not be the servant of ideology, because ideology assumes answers, but science asks questions. The other is that attempts to make science serve ideology will merely make science impotent without assuring that only desired questions are investigated. I am stating simply that we should not control the direction of science and, moreover, that we cannot do so with any precision.

Before trying to substantiate these arguments I must make a crucial distinction. The arguments pertain to basic scientific research, not to the technological applications of science. As we go from the fundamental to the applied, my arguments fall away. There is every reason why technology should and must serve specific needs. Conversely, there are many technological possibilities that ought to be restrained.

To return to basic research, let me first consider the danger of restricting types of investigation because their outcome could be disruptive to society. There are three aspects of danger. One is the fallacy that you can predict what society will be like even in the near future. To say, for instance, that it would be bad for Americans to live longer assumes that the birth rate will stay near where it is. But what happens if the birth rate falls even lower than it is now in the United States and also stabilizes elsewhere in the world? We might welcome a readjustment of the life span. In any case, we have built a world around a given human life span; we could certainly adjust to a longer span and it would be hard to predict whether in the long run the results would be better or worse.

In a general form, I would call this argument for restricting research the Error of Futurism. The Futurist believes that the present holds enough readable clues about the future to provide a good basis for prediction. I doubt this assumption; to think that the data of today can be analyzed well enough to predict the future with any accuracy seems nonsensical to me.

The second danger in restricting areas of scientific investigation is more crucial: although we often worry most about keeping society stable, in fact societies need certain kinds of upheaval and renewal to stay vital. The new ideas and insights of science, much as we may fight against them, provide an important part of the renewal process that maintains the fascination of life. Freedom is the range of opportunities available to an individual—the more he has to choose from, the freer his choice. Science creates freedom by widening our range of understanding and therefore the possibilities from which we can choose.

Finally, attempts to dictate scientific limits on political or social considerations have another disastrous implication. Scientific orthodoxy is usually dictated by the state when its leaders fear that truths could undermine their power. Their repressive dicta are interpreted by the citizens as an admission of the leaders' insecurity, and may thus lead to unrest requiring further repression. A social system that leaves science free to explore, and encourages scientific discoveries rather than trying to make science serve it by producing the truths necessary for its stability, transmits to the members of that society strength, not fear, and can endure.

The other argument I mentioned in opposing imposition of orthodoxy on science is the practical impossibility of stopping selected areas of research. Take aging as a prime example. It is one of the mystery areas of modern biology. The questions are clear. Why do we get older? Why do organ systems slowly fail? Why does one species of animal live three years and another live for one hundred? These seemingly straightforward questions are unapproachably difficult for modern biology. Not only can we not understand events that occur over years, but we even have difficulty understanding questions about events that require minutes to transpire. In fact, molecular biologists are really only experts in the millisecond range of time. Such times are those of chemical reactions; to understand events in longer time frames probably requires knowing how individual reactions are integrated to produce clocks that measure time in seconds to years. Clues to the great mysteries of biology—memory, aging, and differentiation—lie in understanding how biological systems tell time.

There are a few hints about where answers to the puzzle of aging might be found, but they are only the vaguest suggestions. In such an area of science, history tells us that successes are likely to come from unpredictable directions. A scientist working on vitamins or viruses or even plants is just as likely to find a clue to the problem of aging as is a scientist working on the problem directly. In fact, someone outside the field is more likely to make a revolutionary discovery than someone inside the field.

Another example will help to show the generality of this contention. Imagine that we were living at the turn of the last century and had wanted to help medical diagnosis by devising a method of seeing into the insides of people. We would probably have decided to fund medical scientists to learn how to use bright lights to see through patients' skin. Little would we have guessed that the solution would be revolutionary and would come, not from a medical research scientist, but from a physicist who would discover a new form of radiation, X-rays.[3]

Major breakthroughs cannot be programmed. They come from people and areas of research that are not predictable. So if you wanted to cut off an area of fundamental research, how would you be able to devise the controls? I contend that it would be impossible. You could close the National Institute of Aging Research, but I doubt that any major advance in that field *could* be prevented. Only the shutdown of all scientific research can guarantee such an outcome.

Although they would fail to produce their desired result, attempts explicitly to control the directions of basic research would hardly be benign. Instead disruption and demoralization would follow from attempts to determine when a scientist was doing work in approved directions and when he was not. Creative people would shun whole areas of science if they knew that in those areas their creativity would be channeled, judged, and limited. The net effect of constraining biologists to approved lines of investigation would be to degrade the effectiveness of the whole science of biology.

Put this way, the penalty for trying to control lines of investigation seems to me greater than any conceivable benefit. I conclude that society can choose to have either more science or less science, but choosing *which* science to have is not a feasible alternative.

I must repeat a qualification on this broad generalization: the less basic the research area in which controls are imposed, the less general disruption will be caused by the controls. The development of a specific sweetener, pesticide, or weapon could be prevented with little generalized effect.

While it seems necessary that scientific research be free of overt restraints, it would be naive to think that science is not directed at all. Many crucial decisions are made about general directions of science, usually by the control of available resources. Again, the formula I used before is applicable: the less basic the area of research, the easier it is to target the problems. A fallacy behind some of the hopes for the "War on Cancer" was the assumption that the problems to be solved were sufficiently well defined to allow a targeted, applied approach to the disease. Some problems could be defined and those were appropriate targets for a war on the disease, but the deeper mysteries of cancer are so close to the frontiers of biology that targets are hard to discern.

Unselective Limits to Science

I have painted a picture of inexorable, uncontrollable development of basic scientific knowledge. The response of many people to such a vision might well be "If we can't put any controls on research maybe we shouldn't have any research." I see the rationale in this criticism because it is conceivable that the rate of accretion of knowledge could become so high that a brake might be needed. A way exists to produce a slowdown, that is, by controlling the overall availability of resources. A nonselective brake on fundamental biology would decrease productivity without the disruptive and dangerous effects of trying to halt one area and advance another.

Because such a brake was applied in the late 1960s, today we have less basic research, measured in constant dollars spent on it, than we had then. As a result of the slowdown, the danger seen by many is not that we have too much basic research, but rather that we are living on our intellectual capital and that an infusion of funds into basic areas of research is needed before these scientific resources are exhausted. It must also be realized that our commitment to the solution of problems like cancer requires that we develop much more basic knowledge.

I may seem to have strayed from my topic of recombinant DNA research, but I think not. In 1977, there was a draft bill in the United States Senate to set up a National Commission on Recombinant DNA Research. The charge to this commission discussed mainly questions of safety, but at the end of the bill questions of ethics and morality entered. The bill was a clear invitation to begin the process of deciding what research shall be allowed and what research prevented. The battle to stop the creation of this commission was waged by scientists, university presidents, and other concerned individuals across the country. I believe that the long-term success or failure of these efforts will determine whether America continues to have a tradition of free inquiry into matters of science or falls under the fist of orthodoxy.

To finish this discussion, let me quote from the most eloquent analyst of contemporary biology, Lewis Thomas, the president of the Memorial Sloan-Kettering Cancer Center in New York. Writing in the *New England Journal of Medicine* about the recombinant DNA issue, Dr. Thomas ended with this analysis of the role of scientific research in the life of the mind:

> Is there something fundamentally unnatural, or intrinsically wrong, or hazardous for the species, in the ambition that drives us all to reach a comprehensive understanding of nature, including ourselves? I cannot believe it. It would seem to me a more unnatural thing, and more of an offense against nature, for us to come on the same scene endowed as we are with curiosity, filled to overbrimming as we are with questions, and naturally talented as we are for the asking of clear questions, and then for us to do nothing about it, or worse, to try to suppress the questions. This is the greater danger for our species, to try to pretend that we are another kind of animal, that we do not need to satisfy our curiosity, exploration, and experimentation, and that the human mind can rise above its ignorance by simply asserting that there are things it has no need to know.[4]

REFERENCES

[1]R. L. Sinsheimer, "Inquiry into Inquiry: Two Opposing Views," *The Hasting Center Report*, 6 (4) (1976): 18.

[2]L. R. Graham, "Political Ideology and Genetic Theory: Russia and Germany in the 1920's," *The Hasting Center Report*, 7 (5) (1977): 30–39.

[3]Dr. Mahlon Hoagland first suggested to me this example.

[4]P. 328 in *New England Journal of Medicine*, 296: 324–328.

LYNN WHITE, JR.

Science and the Sense of Self: The Medieval Background of a Modern Confrontation

I

ABOUT THE YEAR 1200, Gottfried of Strasbourg composed one of the masterpieces of German literature, his *Tristan und Isolde*. Poets did not yet write in the vernacular to be read, but rather to be heard. One must picture Gottfried reciting, or chanting, his story in a baronial hall, before an audience of nobles, their ladies, and clergy.

King Mark is belatedly informed by a tattletale that his wife Isolde and his nephew Tristan are having an affair. In orderly feudal style he summons his chief vassals to council, and they advise him that the queen must clear her good name by ordeal, that is, by walking, barefoot but unscathed, upon white-hot plowshares. Since, of course, she is guilty, she and Tristan (who has discreetly gone underground) realize that she cannot pass the test; so they resort to a sublime stratagem. Tristan comes to the place of the ordeal thoroughly disguised as a ragged beggar and places himself along the path that Isolde will take on her way to the ordeal. As she passes him, she feigns to swoon, and the official in charge commands the husky bystander to bear her to the spot of the ordeal. Regaining her composure, Isolde swears on all available holy relics that she has never lain in the arms of any man, save only of her husband King Mark and of this filthy fellow who has just carried her a few yards. She then treads the burning irons without blister or blemish. No doubt Gottfried's listeners—certainly some of them clergy—howled with glee at the spectacle of a smart woman using ambiguous words to manipulate divine justice, in Gottfried's words, "like any windswept weathervane."

Was Gottfried encouraging skepticism of religion? Quite the contrary. He was using his satirical genius to encourage the novel brand of Christian Latin theology that had been emerging for at least a century. We are only beginning to plumb the meaning of this development for the altering forms of Western culture. It centered on the three Persons of the Trinity and their historical activity. The figure of Christ became intensely humanized, resulting in a piety typified by St. Francis. The Holy Spirit was the subject of deep contemplation that reached notable proportions in the early twelfth century in the work of a Rhineland Benedictine abbot, Rupert of Deutz, and led by the end of the century to the revolutionary new concept of history propounded by the Cistercian Abbot Joachim of Fiore in Calabria: a vision of human temporal destiny as divided into

Ages of the Father, the Son, and the Holy Spirit, that eventually became, in secularized form, the conviction of the inevitability of moral progress—despite temporary and partial relapses—held by Vico, Hegel, and Marx. But what concerned Gottfried—and concerns us here—is the new theology's increased emphasis on the creation and the majesty of the "one God, the father almighty, maker of heaven and earth, and of all things visible and invisible."

This Creator God was beyond all human comprehension, unimaginably transcendent. Ordinarily he administered his creation by delegation to "secondary causes," agents, notably man. He had endowed human beings with two great gifts, freedom of the will and the capacities of intellect. Increasingly it seemed a bit blasphemous to inform such a God that he must show up at ten o'clock next Tuesday morning in a certain town square to perform a miracle if a woman charged with adultery were in fact innocent. After all, the mind itself is one of God's greatest miracles. Couldn't sensible human judges use it to settle such a problem according to logic and the rules of evidence without inconveniencing God?

In the late eleventh century Roman law had been revived at Bologna and legal faculties were springing up in all the budding universities of the Continent. In the 1140s Gratian codified canon law so satisfactorily that the Church felt no need to repeat the process until the early twentieth century. As for England: within fifty years after Gottfried composed *Tristan*, Bracton produced his treatise on the common law which remained standard until Blackstone's *Commentaries* of the 1760s. In Gottfried's day the new theology was encouraging the growth of rational legal processes, while enthusiasm for these latter helped to implement the new theology. In 1215 the Fourth Lateran Council, under the leadership of Pope Innocent III, abolished ordeals by forbidding priests to invoke supernatural sanction for them. Gottfried's mirth helped to shape the public opinion needed for acceptance of this edict.

No one, of course, doubted that God could, and occasionally did, intervene supernaturally in the course of human and natural affairs. Indeed, every celebration of the Mass was a repetition of the central miracle of Christ's incarnation. Yet Europeans were coming to feel increasingly that, by and large, both men and nature normally operate according to laws which embody their Creator's will for them. People believed that human beings do not make laws: they discover them. It appears that the first person ever to use the term "a law of nature" (lex naturae) in reference not to man but to physical nature was Friar Roger Bacon in the 1260s, and he was clearly conscious of its analogy with human law. Law, then, is inherent in God's purpose for all his creatures. It follows that he cannot be expected to tamper with it frequently in special instances. God is chiefly praised by the perfection with which his creatures exist according to the laws that he has established for and in them.

This shift in the tone of Latin Christian spirituality in the twelfth and early thirteenth centuries was, of course, simply an intensification of part of the Jewish heritage of Christianity. In Greco-Roman religion there was no creation myth; the gods were powerful but not omnipotent; time was cyclical and repetitive; with the decay of each cycle the gods themselves fell, and the whole historical process started over again. Judaism may well have originated in this general Mediterranean pattern, but, thanks to the most remarkable series of religious leaders of whom we have record, it evolved in unprecedented ways. Yahweh, at first the tribal deity of Israel, became the sole God. Abiding before and beyond time and space, totally omnipotent, by an act of will he had created everything from nothing.

Moreover, to the pagan ancients, while at the top of the hierarchy of gods stood the squabbling and mutually frustrating commune on Olympus, there was an infinite downward gradation of godlings, ending in the spirits of trees, rocks and springs—the "spirits of the place." This animism penetrated all of nature and profoundly influenced attitudes towards nature even among Greek scientists. The Judaism of Roman times, however, with its belief in an all-powerful and absolutely transcendent God, had desacralized nature.

So great is the prestige of the Greeks even today that it is hard for a twentieth-century mind, inheriting the medieval idea of natural law as applied to things physical, to grasp the whimsicality and the blind, irrational necessitarianism that were compounded, but not blended, in the pagan view of the world. When Christianity, a Jewish apocalyptic sect, broke with Judaism and took the gospel to the gentiles, it challenged the entire cosmology of the Greco-Roman cultures largely in contemporary Jewish terms. Since they awaited the end of the world momentarily, the early Christians had little interest in science, the study of something that would shortly turn to ashes. Only God and the soul deserved contemplation because only they had any permanence. Nevertheless, one of the chief reasons why the new religion swept the urban population—especially the lower classes—of the Roman Empire so swiftly into its fold was that it provided them for the first time with a clear picture of how the world had come into being, how it would end, and how the Christian believer might better his personal destiny within this cosmic frame. Above all, the Christian was no longer worried about constantly placating, or avoiding offense to, the cloud of minor gods and spirits that buzzed about the head of a pagan.

As the etymology of words like *civil* and *civilization* shows, the sophisticated culture of the ancient world was an affair of cities, and Christianity soon dominated the cities and that higher culture. But because of low rates of agricultural surplus production, the masses of people lived in the pagi, the country districts ruled by the cities. For centuries few of these pagani were much touched by the new world-view, and among them the old animism continued. Networks of rural parishes began to be established in Italy in the later fifth century, and north of the Alps in the sixth, but even in the tenth century they seem to have been lacking in some areas. Moreover, the conversion of Celts and Germans, Slavs, Magyars, Scandinavians, and Balts went slowly; in the 1070s human sacrifices were still celebrated at the great Norse shrine of Uppsala, and the Lithuanians clung to their ancient gods until 1386. Theoretically the Christian cult of saints was unrelated to the pagan cult of local spirits, but in practice it doubtless at times masked animistic ways of thinking. Generation after generation the Church kept hammering against pagan survivals and any hint of the sacred in natural objects, save only the bread and wine of the eucharist. Its drive in the twelfth century to stress the omnipotence of God seemed to assure the final victory of this basic Judeo-Christian view of reality over the residue of paganism.

Yet at precisely this moment a strong countercurrent threatened to revive, not ancient paganism as a religion, but—more dangerous—the underlying presuppositions of much of that paganism. The church of the first Christian millennium had made a truce, if not a peace, with Plato in his neo-Platonic version which has no great concern for the concrete phenomena of nature. Aristotle, however, had been largely ignored, save for part of his logical works. Quite suddenly the situation changed. Perhaps the new scientific impulse was related to an indigenous Western theological insight that we have a religious duty to use the gift of the intellect to try to understand how the omnipotent Creator has chosen to order his cosmos. In the twelfth and thirteenth centuries great numbers of scientific works and books of science-oriented philosophy were translated not only from Greek but also from Arabic. Much of Arabic science was based on Greek material, even when it advanced beyond it; and the Arabic commentators proved a great stimulus.

In the 1150s a Norman Sicilian, Henry Aristippus of Catania, translated Plato's *Meno* and *Phaedo* from the Greek. But the public did not want the vague, mythopoeic intellectual commodity that Plato offered. Aristippus's translations had no sale, and Plato had to wait for Ficino and Medicean Florence. On the other hand, the later twelfth century began to discover Aristotle, the polymath who had written on everything from rhetoric to zoology, politics to metaphysics, and who offered not only generalizations but also a wealth of specificities. He was just what many of Europe's intellectuals wanted, and translations of his works swept the book market.

Unfortunately, however, Aristotle was a Greek pagan. He did not believe in a creation, but rather that the world is eternal. He did not believe in the immortality of the individual soul. Above all, he believed that everything, including the gods, is bound by inherent necessity: in his universe there is no freedom, no option, consequently—at the deepest level—no virtue and no vice. Understandably some of the clergy began to worry. Their first reaction was to prohibit his books, or at least public lectures on the more controversial of them. But Aristotle could be neither repressed nor ignored; he met too many contemporary needs. So at last Thomas Aquinas started out to make a Christian of Aristotle.

It very nearly worked: Thomas had the greatest genius for systematization in the entire history of philosophy. Naturally at points of frontal collision, like creation and personal immortality, the epistemology of revelation took precedence over "reason" in the form of Aristotelian logic. It is astonishing, however, how seldom such confrontations occurred. The whole thrust of Thomas was to show not only how beautifully logical revealed dogmas can be, but even how logically necessary they are.

The neo-Augustinians who had fostered the new theology of the twelfth and earlier thirteenth centuries were far from stupid. If sweet reason could provide so broad a foundation for Christian faith, the need for revelation was being called in question. To them the Thomistic–Aristotelian synthesis was a Trojan horse of resurgent paganism.

Thomas had made one tactical blunder: he had adopted Aristotle's principle of economy. For example, nature abhors a vacuum because it would be a waste of space; there can be only one world because another would be exactly like the first, and redundant; since the circle is the most economical rotund form, all celestial bodies move in circular orbits; because the world contains all that is necessary, no new forms (species) can be created. Thomas agreed. He could not imagine a lavish, spendthrift Creator.

In 1277, three years after Thomas's death, the defenders of the dogma of God's absolute omnipotence struck back. Étienne Tempier, bishop of Paris, pronounced automatic excommunication against anyone teaching any of 219 condemned propositions, most of them clearly Aristotelian and sometimes Thomistic. If God wishes to create a vacuum, or create a plurality of worlds, or move a heavenly body in a pattern other than a circle, or produce new species, that is within his options, which are limitless. Whether we puny creatures can ever detect vacua or other worlds or noncircular orbits, or find an entirely new animal, is irrelevant; nothing can be excluded by a Christian as beyond God's capacity. The message from the bishop to the professors of the University of Paris was loud and clear: Quit telling God what he can and cannot do.

Considering how powerful it was, and how ruthless it was in stamping out heresy among the common people—not only through the gruesome procedures of the Inquisition but also most notably by the hideous slaughter of the Albigensian Crusade—the thirteenth century Church was remarkably restrained in dealing with religiously deviant intellectuals who simply talked to other intellectuals. After all, at least in northern Europe, most university men were clerics, that is, "in the club." Moreover, their education in dialectic trained them to use logical discussion as a research method, and often a scholar in debate would uphold a proposition that he did not personally espouse. Vagaries generally were overlooked provided that one made no effort to share them with people ignorant of the rules of the scholastic game. It was not until the vitality of scholastic education ebbed in the sixteenth century, and the Protestant secession frayed tempers, that ecclesiastical tolerance of wayward intellectuals was seriously eroded. The only casualty connected with Tempier's edict was a particularly obstreperous young professor, Siger of Brabant, who had already felt the bishop's hot breath in 1276 and had fled Paris to the papal court, then in Orvieto, whose judgments were generally more lenient than those of local authorities. There he was put into mild protective custody awaiting hearings. Since Dante places Siger in Paradise in the circle of the great theologians, Italian opinion was presumably somewhat favorable to him. Unfortunately, however, his custodian went insane and stabbed him to death.

Scientists in modern America, fearful of infringement of their freedom by nonscientific outsiders, may not be expected to view Tempier's sweeping decree of condemnation charitably. Yet scientists as a group seem seldom to be more prone to reexamine their axioms than are less intellectually confident people. Respect for ancient written authority was immense during the Middle Ages, and the recent discovery by the West of the vast and fascinating corpus of Aristotle's works quickly made him—in Dante's phrase—"the master of those who know." On the basis of a Christian dogma, the bishop of Paris demanded that natural philosophers start thinking about nature in non-Greek terms.

Nothing more salutary could have happened to Western science. All over Europe during the late thirteenth and fourteenth centuries natural philosophers, like the lawyers a bit earlier, felt required to take a fresh look at their thought processes, semantics, and the criteria of validity in evidence that led them far beyond Aristotle in the direction of modern logic. They greatly narrowed the concept of causality, and pushed strongly towards a purely empirical examination of natural phenomena. While their experiments were usually—although by no means always—"thought experiments" rather than laboratory tests, the results were startlingly novel in optics and mechanics particularly. In many ways they paved the road toward the age of Galileo and Newton.

II

While, under the prod of theological dogma, Western scientists were establishing an entirely un-Greek concept of natural law and were also abolishing classical notions limiting what might be possible in nature, Christian piety was putting a new emphasis on the observation of matter.

The core of Christianity is the belief that the totally omnipotent and transcendent God put aside both transcendence and power to become a human being precisely like ourselves in order to free us from the sins arising from misuse of the will and the intellect. At the end of his work of creation, God had looked at his material world and pronounced it good. By his incarnation God validated this goodness of matter. It is much easier to believe in a majestic, transcendent god than in one who is a carpenter, born in a stable and dying in agony on a scaffold. The art of the first Christian millennium tended to evade both Bethlehem and Calvary; Christ was normally shown as the majestic judge at the last day.

During the twelfth century attitudes changed in the Latin West. We do not know why. Clearly, however, this mutation of piety had something to do with the great waves of pilgrims in the eleventh and twelfth centuries who swept eastward to experience for themselves the dust and heat and thirst of the roads of Palestine, and who walked the Via Dolorosa on bloody knees. The greatest preacher of his age, St. Bernard, cried "I who am a man am talking to you who are men about a *man*!" In an Italian village St. Francis at Christmas first put a feed-trough and an ox and ass together in a barn to show the peasants just how it was when God became a baby. On the head of the crucified Christ the old regal crown was replaced by a crown of thorns, and the formerly passionless body now writhed in pain. Latin Christendom, like St. Thomas the Apostle, thrust its hand into the spear-wound in the side of Christ. Piety demanded empirical experience.

Nowhere was this more the case than with the eucharist, since every Mass was a repetition of both Bethlehem and Calvary. The Church had accepted the dogma of Christ's physical presence in the consecrated bread and wine of the altar, but had felt no need either to ask or to say exactly how this could be: to this day the Eastern Churches are content to let mystery be mysterious. Not so the West. In 1215 the Fourth Lateran Council, which had banned ordeals, also felt cultural pressures to explain what happened to the bread and wine in the priest's hands. It formulated the dogma of transubstantiation in terms of the best theory of matter then available, the distinction between substance and transient qualities. The substance of water remains unchanged even though it can be made to appear as a drop, a puff of steam, or an icicle. Lateran IV assumed that it was logical that in supernatural circumstances the substance of matter may be changed without affecting its external characteristics. Thus the bread and wine of the altar continue to look like bread and wine even though in fact they have become the body and blood of the incarnate God.

The interest of this for the history of science is not the validity of the explanation, but rather the unprecedented quality of the public mentality which demanded that a material mystery be explained. It appears that most of the faithful were satisfied.

The result was an explosive development in the thirteenth century of a new sort of eucharistic piety. Hitherto, consecrated wafers had been reserved for peoples' veneration between Masses in enameled or jeweled containers. But now Europeans wanted to see for themselves the body of the Christ whom they worshiped; so transparent monstrances carved of rock crystal were developed. Since these were costly, ingenious glassmakers developed a clear glass that did as well for the empirics of poorer parishes. The enthusiasm for actually seeing what one adored quickly spread to relics. Formerly they had been placed in decorated opaque reliquaries. Now reliquaries likewise were made transparent. Holy bones were everywhere visible. For very small relics, like splinters of the true cross or thorns from the crown of thorns, it was found by the increasingly skillful glass cutters that a convex surface made them more visible. (One casual spinoff was the invention of eyeglasses in lower Tuscany in the 1280s.) By the end of the thirteenth century it seemed that Western Europe, by such intentness on the matter of its deity, had come close to deifying matter and making the understanding of matter a prime spiritual concern. To us, looking back, a new nonpagan but dangerous sacralization of matter appears to emerge.

Matter is understood most competently with measurement and numbers. In the late tenth century an English illumination shows for the first time the hand of the Creator God holding an architect's square, compasses, and scales. This iconographic tradition continued into modern times, culminating in William Blake's "Ancient of Days": a mighty bearded figure leaning down over the abyss of nothingness and, with immense compasses, drawing the circle of the cosmos.

The mathematics of the early Middle Ages in the West was fairly rudimentary, but the great movement of translation quickly remedied that. In 1100 only scraps of Euclid's *Elements* were available in Latin. By 1200 there were six complete versions, one from Greek, the rest from Arabic. While Aristotle's science was more qualitative than quantitative, other works brought westward the mathematics not only of Islam but also of India. The first great mathematical book ever written in Latin was published in 1202 by Leonardo Fibonacci, a Pisan merchant who had been brought up in Algeria, spoke Arabic fluently, and had studied Islamic science with Muslim scholars. His declared intent was to teach his fellow Europeans how to calculate "the way the Indians do," but in addition to Indic arithmetic he transferred to the West much of Arabic mathematics. Others were transplanting Sanskrit trigonometry. This was picked up and developed in the early fourteenth century by Franciscan friars at Merton College in Oxford and by Rabbi Levi ben Gerson in Provence. His Hebrew writings were soon put into Latin—two of the translations were dedicated to Pope Clement VI—and these fused with the stream from Merton to produce a more advanced variety of trigonometry. By the middle of the fourteenth century, Europe had garnered the mathematics of all Eurasia and was seizing leadership.

Traditionally the seven liberal arts had been divided into the trivium (grammar, rhetoric, and logic) and the quadrivium (arithmetic, geometry, astronomy, and music, conceived as the study of acoustical proportions). The first dealt with verbal methods of analysis, the second with measurement and calculation. In the fourteenth century both natural scientists and theologians were thoroughly grounded in the full range of the liberal arts and shared the same realm of discourse; when they disagreed they were at least talking to each other and not past each other. Indeed, a high proportion of scientists were also technically trained theologians.

As the amazing intellectual dialogue started by the 1277 condemnations grew and matured, a curious, and in many ways ominous, consensus began to appear. The effort to achieve a total vision of truth, that had been the goal of the scholastics of the twelfth and most of the thirteenth centuries, began to seem futile. With the rigorous tightening of standards of proof both in natural philosophy and in theology, there came increased confidence in the validity of results achieved quadrivially and a decline in conviction of the cogency of trivial argument. The scientists tended to narrow their research methods to the mathematical, and their topics to the physical. The theologians of the fourteenth century, on their part, widely concluded that while logic might help to clarify belief, the Christian faith rested on revelation, including the continuing revelation reflected in the Church's tradition, and was indemonstrable by reason. Increasingly, in theology, explicit paradox, the complexio oppositorum, or coincidence of opposites—a mode unacceptable in mathematics—became the most satisfying form of expression. Logic in the fourteenth century achieved amazing sophistication, and in the process destroyed its utility for either science or religion.

In such circumstances, how do people think successfully about problems of value? Among his other achievements, Bishop Nicole Oresme, the greatest French intellectual of the later fourteenth century, invented the graph. Once he tried to graph beauty. It didn't work. Natural reason had become too identified with its most precise expression, mathematics, to cope with a vivid but diffuse experience like that of beauty. The inherent limitations of reason were now recognized so clearly that the whole realm of qualitative values—beauty, goodness, even truth, when the nature of that truth was not precisely calculable—found refuge in theology with its epistemology of revelation. The essence of C. P. Snow's "two cultures" is to be found in Europe 600 years ago. All that was "holy," all the intangible, unmeasurable values, had been purged from nature, partly by pressure of the Jewish doctrine of divine transcendence and omnipotence transmitted through Latin Christianity, but likewise by the new Western piety towards material substance and its reflection in more powerful mathematics.

III

This schism within Western culture between ways of coping with the tangible and the intangible has been the more traumatic in recent generations because no other civilization has given greater valuation to the internal, "subjective" experiences of the individual. The Sufis of Islam, the T'aoist eccentrics of traditional China, have been seminal in society but seem seldom to have been close to its center. In the Occident, by contrast, individualism—save in the totalitarian regimes of our century—has normally been one of the major ideal patterns of personality. Its roots go deep.

Anthropologists sometimes distinguish between "shame cultures" in which the sanction against evil-doing is public ostracism, and "guilt cultures" in which a concealed misdeed may cause the culprit profound distress. In pagan Antiquity ethics was a purely philosophical topic; since mythology showed the gods committing every crime from murder to pederasty, they were scarcely in a position to punish human beings for anything but sacrilege or blasphemy. From the time of Moses, however, Yahweh was intensely concerned about the quality of conduct on the part not only of his chosen people as a whole but also of individuals. From his eyes and his ultimate justice, nothing could be hid. Christianity inherited this conduct-oriented deity from its mother religion, and challenged the shame-culture of Antiquity with its own guilt-culture.

Confession and penance became a major sacrament. At first, confession seems often to have been public, but not infrequently this created scandal affecting innocent persons; so confession rapidly became not merely private but also confidential. By the sixth century it was established that "the ear of the priest is the ear of God."

As the network of parishes spread out of the cities to cover the countryside during the early Middle Ages, so did the custom of confessing. In Ireland as early as the sixth century so-called Penitentials began to be written; they were handbooks for the evaluation of sins and the definition of appropriate penances. For some three centuries Irish missionaries flooded the mainland as far as Eastern Germany, Central Italy, and Northern Yugoslavia, spreading the use of such books. Not only the upper classes but even the humblest peasants came at length to kneel by the chair of a priest, whisper their derelictions to him, and talk to him about how to do better. In 1215, Lateran IV (how often, almost as in a rondo, we find ourselves returning to that great conclave, so perceptive of the needs and wishes of its people!) decreed that every Latin Christian must confess at least once a year. No other society has ever systematized the examination of conscience so thoroughly. And since knowing ourselves is the cornerstone of individualism, no other society has so motivated the development of self-conscious individuals at every social level.

In the same period a change in domestic architecture was supporting the growth of individualism. Roman methods of heating buildings worked reasonably well for the Mediterranean climate, but not for the rainy, variable, and often cold regions of Northern Europe. There, especially in winter, life in baronial mansion or peasant cottage centered around a central fireplace with a covered louvre in the roof to permit escape of smoke and sparks. Everyone ate, spent their spare time, and often slept there. Society was hierarchical, but the levels of the hierarchy knew each other well, if only from forced adjacency.

During the ninth century the chimney and mantled fireplace appear, first at the abbey of St. Gall in Switzerland. A well-designed chimney flue drew the smoke out through the wall while radiating considerable heat into the room. It was safer than the central fireplace because sparks were more easily controlled. It was more economical because small fires with chimneys running up the walls could be built in small rooms in multistoried masonry buildings, and these gradually replaced the big fire in a great hall the roof of which was necessarily under the sky. By the eleventh century chimneys and mantled fireplaces were normal in noble residences and by the end of the twelfth were common in peasant houses as well. The lord and lady began to spend less time in the great hall and to seek privacy in their withdrawing rooms. Soon the chief retainers of the household also secured warm private quarters, and the upstairs–downstairs syndrome appeared as the lesser dependents were isolated from their betters. In the 1370s William Langland lamented that nowadays "every rich man eats by himself in a private parlor to be rid of poor men, or in a chamber with a chimney, and leaves the great hall." The admirable scholar who has done most work on the history of the chimney flue has remarked that it doubtless did more for the art of love than all the troubadors combined. And one may add that by implementing privacy it provided an essential milieu for the growth of the western ideal of the idiosyncratic individual.

IV

In our own time, a high proportion of ultraindividualists who are not themselves scientists profess alienation from science even when they continue to enjoy its by-products. How did their spiritual antecedents of the later Middle Ages view the increasingly mathematized and matter-oriented science of their day? Quite cordially, it would seem. This demands explanation, at least to us of the later twentieth century.

Astrologers had thrived in the Hellenistic and Roman worlds. During the early Middle Ages their art was still practiced marginally, but the Church considered it a pagan superstition and fought it with remarkable success. Then the tide turned. The more carefully we probe the datings, sequence, and substance of the twelfth-century Latin translations from Greek and Arabic, the clearer it becomes that astrology was central to the first surge of scientific interest in the West. Much of medicine, mathematics, astronomy, botany, and the like traveled to Europe on the magic carpet of astrology. It was not until towards 1200, when the Aristotelian corpus became central to the thinking of Latin scientists, that astrology ceased to be pivotal. It remained, however, integral to the scientific movement and was more closely related to popular concerns than any other branch of science.

Sun and Moon obviously affect human affairs by guiding the seasons and the tides. It seemed to many people probable that the other five visible planets were likewise influential, even upon individual destinies. In its most elaborate form, astrology was a complex discipline involving careful observations of transits of planets and elaborate mathematical processes.

During the early Middle Ages when almost nothing was written in the West on astrology, intellectuals had great enthusiasm for calendrical treatises on the dating of Easter that are in no way astrological but which nevertheless seem psychologically analogous to astrology. Between 526 and 1003 no fewer than eighty such books were produced. To understand the reason, one must put oneself into the boots of a Benedictine monk living in one of the great abbeys of that age.

These monasteries were the most productive centers of learning in Europe. They were, however, communities withdrawn from the world and primarily devoted to the perpetual praise of God. Ideally they were anticipations of the Heavenly Jerusalem where, as St. Augustine said, "our sole duty will be to cry 'Alleluia!' " Life and prayer were corporate and followed the majestic cycle of the Christian year. The personal fulfillment of a monk consisted in his sharing the solemn ecstasy of the orderly succession of liturgies, the choreography of the religious seasons. On a great festival the high altar was a sea of lights, and waves of incense and song, penitential or joyful, rose toward the Creator enthroned beyond his creation. The supreme festival was Easter, yet it was also the variable dominating everything from Septuagesima until Trinity Sunday: a span of four months. Calendrical treatises told a monk how, by calculating solar and lunar motions, he could worship rightly and thus achieve his destiny. They provided his nexus with the visible heavens and with the extracosmic bosom of Abraham toward which he strove.

By about 1100, the intellectual dominance of the Benedictines was ending. The liveliest learning was now to be found usually in cathedral schools, around which clustered freelance lecturers supporting themselves on student fees. European society was becoming much more intricate than it had been. Life was still largely corporate, but the variety of corporate bodies greatly increased: city governments run by burghers; vigorous guilds of workmen or merchants; confraternities of devout laymen; nascent universities, and the like—there were many novel contexts for the growth of a more intense individualism than had earlier been common. Although heresy was spreading, not many people broke with the Church; yet many seemed to be looking for a new nexus between the cosmos and their personal destinies. Astrology, long dormant, sprang to life again because, by exact mathematical means, it purported to unveil the detailed connections between the macrocosm of the celestial spheres and the microcosm of each man. To the idiosyncratic personalities of the new age, study of the gyrations of the seven planets gave the same relief from anxiety that the Benedictines had found in calculating the date of Easter. Fear and hope thus provided a new market, in a new social situation, for Latin translations of Greek and Arabic texts hitherto disregarded in the West. In the face of so novel and pressing a psychological need, the Church gradually reduced its traditional hostility toward astrology.

Very quickly astrology became an essential part of medical practice, and remained so until deep into the seventeenth century. Especially in treating a wealthy patient, the physician was compelled by medical ethics to cast a horoscope, and this demanded not only intricate calculations but often new observations of transits, since the available tables of transits had to be rectified to correspond to the meridian of the patient. Each part of the body was related to a zodiacal sign, and the treatment—by drugs, diet, bleeding, enemas, or the like—depended on how the individual's personal mixture of "humors"—hot, cold, moist, and dry—fitted his astrological situation. During the later Middle Ages some of Europe's ablest minds were professionally engaged in medical astrology, and in no other age have physicians been more widely respected or more amply paid. It is doubtful that their ministrations cured many who would not otherwise have survived, but their fusion of mathematics, astronomy, and medicine provided a consolation, a sense of the individual's connection with the wheeling spheres of the heavens, a confidence that all was right with the world, that was clearly worth paying for. By the fourteenth century even popes and cardinals were maintaining medical astrologers in their retinues.

There was, indeed, some criticism that the available observational instruments inherited from Greeks and Muslims were too crude to give an adequately exact base for diagnosis and treatment. Astrologers responded by perfecting their equipment. The greatest difficulty was precise timing of transits: on chilly nights the flow-holes of water clocks tended to coat with ice, thus ruining observations. By 1260 it was recognized that a weight-driven clock might provide the best solution. Despite much work, however, the mechanical clock was not invented until toward 1330. Considering its astrological context, it is not surprising that the first certain example was more than a timepiece; it also showed the motions of Sun and Moon, and the sequence of the tides. One completed in 1364 had 107 wheels and gears, and showed the revolutions of the seven planets; in addition it had, most significantly, an automatic perpetual calendar of all the Church's feasts, both fixed and moveable. The Church quickly saw that these beautiful and complex machines—often equipped with automata of prophets, saints, and angels who paraded at the striking of the hours—could teach the populace the orderliness and majesty of God's universe, and it did much to spread them. It was Nicole Oresme who invented the metaphor of God the clockmaker who had arranged the celestial spheres like a vast mechanical clock in which all the wheels move harmoniously. This was the deity that Gottfried of Strasbourg had had in mind, but the metaphor was not yet available to him.

At times an overly enthusiastic astrologer would attempt to use his art to predict coming events, and this was normally denounced by churchmen as a contradiction of the human freedom to choose. For this same reason such "judicial astrology" was at times perceived as a threat to the self-images of rugged individualists. Otherwise, however, there was little in the state of European science until the early seventeenth century to cause widespread alienation from it.

V

Then, at first slowly, the relation between individual self-awareness and science, which for so long had seemed mutually supportive, began to change. From the early thirteenth century the term "world machine" (machina mundi) had increasingly been used, and in a favorable sense. Europe was coming to admire ingenious machines, and was inventing many new ones. The expression referred, however, not to earthly operations, but to those of the astronomical spheres. Oresme, when he spoke of God as being like a clockmaker, was thinking only of the planetary orbits; with Aristotle and Ptolemy he believed that friction and change were found solely in the sublunar regions.

In the late sixteenth century a spectacular nova exploded, and even non-scientists began to wonder about celestial immutability. In 1610, on the basis of a Dutch invention, Galileo built an improved telescope and found spots on the sun, mountains on the moon, and satellites wheeling around Jupiter. He thus unified celestial and terrestrial physics, but instead of demechanizing the sky, he ended by mechanizing the earth. His younger contemporary, Descartes, included in this mechanization all animals; to him, and to millions who fell under his influence, they were insensible machines who, despite superficial evidences of joy or agony, had neither minds nor feelings. Only human beings had such characteristics; the mechanical res extensa of man's body was connected to the res cogitans of his mind by curious properties of the pineal gland.

Then it became mankind's turn to be mechanized. Confidence in the pineal gland waned. Darwin made man completely part of nature. There should have been nothing surprising about this. If we can believe anthropologists and the historians of non-Western cultures, most people in most times and places have thought of themselves as being such, and have found sustenance in the idea. The trouble among Westerners was that by the twentieth century the late-medieval emphasis on the importance of matter, and on mathematics as the only basis for rational certainty, had created a lifeless and impersonal image of "nature" with which few wanted to be identified; it violated the sense of selfhood among us which similarly is so largely a product of our medieval ancestral experience.

Who will deny that mathematics can be an art beautiful in itself, and that it has proved to be the essential key to many a scientific puzzle? What must be denied is that mathematics, measurement, and quantification provide much insight into the deepest personal problems and experiences of our race: courage and cowardice, affection and hatred, generosity and greed, charm and repulsion, courtesy and boorishness, awe and mockery—and so on without end. About such matters we can only talk to each other, at length, carefully, trivially. We must talk patiently, weighing contradictions, balancing paradoxes, and not expecting to arrive at "laws" but rather at rough consensus that draws on long experience and takes a long view of the road ahead. Quadrivial ways are useful, but those who employ them habitually because they produce good harvests from their special fields must recognize more vividly the severe limitations of their favorite method and the cultural harm caused by abuse of it. They must help to heal the schism between the trivial and quadrivial methods that today threatens them, the scientists, more than it does any other learned group. The rise of a counterculture was inevitable, and it began generations ago. It will continue to grow, and it will threaten the prosperity and freedom of scientists until our scientific community rethinks its professional legacy from the Middle Ages.

LEO MARX

Reflections on the Neo-Romantic Critique of Science

> I have insisted that there is something radically and systematically wrong with our culture, a flaw that lies deeper than any class or race analysis probes and which frustrates our best efforts to achieve wholeness. I am convinced it is our ingrained commitment to the scientific picture of nature that hangs us up.
>
> The scientific style of mind has become the one form of experience our society is willing to dignify as knowledge. It is our reality principle, and as such the governing mystique of urban industrial culture.
>
> Theodore Roszak[1]

I

SERIOUS, WIDESPREAD CRITICISM of science is a relatively recent development in the United States. Until World War II the national faith in the identity of scientific and social progress remained largely unshaken. Most Americans, even after the Great Depression of the 1930s, continued to regard the life-enhancing value of scientific knowledge as self-evident. But since Hiroshima, public anxiety about the consequences of scientific discovery has risen steadily. The nuclear arms race; the polluting and carcinogenic effects of new petrochemicals and other products of science-based industry; the actual and possible uses of electronic devices as instruments of social control; the prominent part played by certain science-based technologies of a particularly revolting kind in the prosecution of the American war in Southeast Asia; the potentialities for genetic engineering created by advances in molecular biology—these are only the more conspicuous causes for the rising public alarm about the results of scientific research. It is now evident that the American belief in the inherently beneficial character of science no longer can be taken for granted.[2] Judging by current discussion of the subject, however, one might infer that the legitimacy of science—by which I do not mean its lawfulness in any narrow sense, but rather its compatibility with accepted standards and purposes—is now being called into question for the first time.

But in Western culture the legitimacy of modern science has been in question since its emergence in the seventeenth century. To be sure, certain major themes in the legacy from the earlier critique of science have since lost their credibility and all but disappeared. The learned clergy, for example, no longer attacks science as a threat to the churches, or as a deflection from the primacy of theological knowledge. Today no responsible clergyman would think of opposing a scientific research project on the ground that the worship of God precludes the study of nature. But it is necessary to add that serious criticism of science based on religious values, though not expressly identified as such, retains an immense appeal. Some of the more effective of the currently popular arguments against science prove, on inspection, to be secularized versions of an essentially religious, or teleological, conception of knowledge.

Another ancient theme in the critique of science which has virtually disappeared is in effect a defense of the older, aristocratic, humanism. On this view, widely held in the age of Pope and Swift, the proper study of mankind is man, not nature. At stake then was the moral instruction of a small, privileged, ruling class for whom scientific education was deemed inappropriate—which is to say, vulgar, unedifying, merely useful. Like the antagonism toward science grounded in theological preconceptions, this avowedly patrician argument is no longer invoked, for obvious reasons, by critics of science.

But the same cannot be said about the general ideas embodied in imaginative literature beginning in the late eighteenth century. On the contrary, most of the themes which figure prominently in the current criticism of science were anticipated by the writers of the romantic era. They called into question the legitimacy of science both as a mode of cognition and as a social institution.[3] To question the legitimacy of science as a mode of cognition means, I assume, to ask whether the conception of reality implicit in the scientific method is adequate to our experience. Is it reliable, coherent, sufficient? Insofar as it is not sufficient, does it mesh with what we know by means of other modes of knowledge? To question the legitimacy of science as an institution is to ask whether the methods (and products) of scientific inquiry are compatible with the expressed and tacit goals of society. Can the technological consequences of scientific discovery be assimilated, for example, to a more just, healthful, and peaceful social order?

Both kinds of question, to repeat, were implicit in the literary response to scientific rationalism and the innovations, technological and social, associated with it. At the core of the romantic reaction, in the well-known formulation of Alfred North Whitehead, was "a protest on behalf of the organic view of nature, and also a protest against the exclusion of value from the essence of matter of fact."[4] Implicit in each of these "protests" is a negative, or potentially negative, answer to the questions about the legitimacy of science raised above. The protest on behalf of the organic view of nature is directed against the presumed epistemological insufficiency of science. Scientific method is thus held to be inadequate to the (unified) nature of nature, which is assumed to be a whole distinct from the sum of its parts, and hence not apprehensible by means of the piecemeal, or analytic, procedures which dominate (normal) scientific inquiry.

The second "protest" identified by Whitehead (against "the exclusion of value from the essence of matter of fact") is applicable to both the cognitive and institutional senses of "science." It means that as a method of knowledge science lends insufficient expression to the distinctively human attributes of reality, those which are properties of mind rather than merely of natural objects. But "exclusion of value" also may be taken as a reference to the negative social and political results of scientific neutrality. It anticipates the now familiar charge that scientists do not assume adequate responsibility for the social consequences of their work. The substantive moral neutrality of natural science as a method of inquiry is not a warrant, in this view, for the morally uncommitted posture of scientists outside their laboratories or classrooms. My point, in any case, is that much of today's criticism of science, including the antagonistic viewpoint widely disseminated by spokesmen for the dissident movement, or counterculture, of the 1960s, may be traced to the double-barreled romantic reaction of European intellectuals which began in the late eighteenth century.

II

The mainstream of the European critique of science entered American literary thought under the auspices of Ralph Waldo Emerson. The English writers who chiefly influenced his thinking on the subject—Wordsworth, Coleridge, Carlyle—had in turn been influenced by the several versions of post-Kantian idealism then being imported into England from Germany. If we accept for the moment the standard, over-simplified handbook view of the spectrum of English literary attitudes toward science as extending from the hostility of Blake at one extreme to the admiration of Shelley at the other, then the writers who were congenial to Emerson must be accounted middle-of-the-roaders. In view of today's assumptions about the antagonism between the "two cultures," in fact, these English moderates would seem, like Goethe, to have been remarkably hospitable toward the claims and prospects of science. Although in one way or another they all recognized the limitations of scientific rationalism, they expressed no serious doubts about the inherent validity of the scientific method. Nor were they frightened by the prospect of the revolution in the conditions of life soon to result from the application of the new science to the fulfillment of economic needs. Their optimism was most directly expressed in repeated assertions about the essential compatibility between scientific and other modes of perception, especially aesthetic or literary, like this well-known statement of Wordsworth's in the *Lyrical Ballads* preface of 1800:

> If the labors of men of science should ever create any material revolution, direct or indirect, in our condition, and in the impressions which we habitually receive, the poet will sleep then no more than at present; he will be ready to follow the steps of the man of science, not only in those general indirect effects, but he will be at his side, carrying sensation into the midst of the objects of the science itself. The remotest discoveries of the chemist, the botanist, or mineralogist will be as proper objects of the poet's art as any upon which it can be employed. . . . If the time should ever come when what is now called science, thus familiarized to men, shall be ready to put on, as it were, a form of flesh and blood, the poet will lend his divine spirit to aid the transfiguration, and will welcome the being thus produced, as a dear and genuine inmate of the household of man.[5]

Behind such optimism about the future collaboration between science and poetry ("poetry" usually taken to represent aesthetic and moral discourse generally) was the assumption that the two modes of perception stand in a potentially complementary relation to each other. Thus at least the majority of scientists were presumed to be operating at the level of the "Understanding." This is the empirical mode of apprehending external reality, based upon sense perception, as set forth in John Locke's "sensational" theory of knowledge. This theory had proven, by Wordsworth's time, to be ideally suited to negotiations, theoretical and practical, with the world of material objects. As Emerson put it, in what came to be recognized as the philosophic manifesto of American transcendentalism, *Nature* (1836), the Understanding is the capacity of mind which "adds, divides, combines, and measures," whereas the Reason, a mythopoeic, analogizing, intuitive mode of perception, "transfers all these lessons [of the empirical Understanding] into its own world of thought, by perceiving the analogy that marries Matter and Mind."[6]

Assertions on the plane of the Understanding are in effect data-bound, and require only literal language, whereas assertions on the plane of Reason transcend the "natural facts," require figurative language, and thus contribute to epochal rearrangements of thought and feeling. Emerson's favorite illustration of this point, following the analogous distinction between the Fancy and the Imagination, is the difference between the practice of a merely fanciful poet and that of a truly imaginative genius, a Virgil, Dante, or Milton, whose work effects a symbolic reconstruction of reality. But Emerson clearly meant the distinction to apply, by analogy, to the work of scientists as well. It is the difference between routine science, which merely elaborates, confirms, and refines an established theoretical structure, and the revolutionary syntheses of a Galileo or a Newton. So far as Emerson's theory of knowledge can be taken as tacit criticism of science, it is directed against inquiry confined to the plane of the empirical Understanding. Besides, he is only questioning the sufficiency, not the reliability, of such knowledge. But the distinction between the two modes opens the way for those critics who would charge science with encouraging a dangerous imbalance on the side of the instrumental, empirical Understanding.

Thomas Carlyle was one of the first writers to invoke a similar distinction between two modes of knowledge as a way of calling into question the legitimacy of modern science as a social institution. In "Signs of the Times" (1829), he locates the governing spirit of the "Age of Machinery" in the empirical philosophy of John Locke. Locke's "whole doctrine," he asserts, "is mechanical, in its aim and origin, its method and its results. It is not a philosophy of the mind; it is a mere discussion concerning the origins of our consciousness, or ideas . . . a genetic history of what we see *in* the mind." By "mechanical" Carlyle refers to Locke's emphasis upon the accumulation and sorting of external data (accomplished by the Understanding), and a minimizing of the active, synthetic, and transformational powers of mind (accomplished by Reason). Because he conceives of the contents of the mind as contingent upon sense experience, upon facts flowing in from the outside, Locke tends to reduce thought to a reflex of the environment. This way of knowing is extremely useful for manipulating physical reality, but it leads to a quietist abdication, or fatalism, with respect to the controlling purposes of man's newly acquired power over nature. "The science of the age . . . is physical, chemical, physiological; in all shapes mechanical," hence the image of a machine best characterizes the dawning era of instrumental reason.

But the "mechanical philosophy," as Carlyle describes it, need not be destructive. On the contrary, it could be advantageous to mankind, and in fact he admits that in certain respects the age is advancing. The trouble is, however, that the advance is grossly unbalanced: while the physical sciences are thriving, the moral and metaphysical sciences are falling into decay. Carlyle's complaint in effect belongs to a later stage in the response to the advance of science which Lynn White has discerned as early as the fourteenth century. As scientists tended "to narrow their research methods to the mathematical, and their topics to the physical," according to White, there was a decline in conviction of the cogency of "trivial" argument, that is an argument in the essentially rhetorical language of the "trivium" as against argument in the essentially mensurative language of the "quadrivium."[7] In Carlyle's view, in any case, scientific rationality is spreading far beyond its proper sphere, and the result is that the culture is permeated by "mechanical" or technological thinking. The age of machinery overvalues those aspects of life which are congenial to the "quadrivial" mode of thought, to use the terminology of the earlier age, which is to say they are quantifiable, calculable, manipulatable. By the same token it downgrades the sphere of the moral, aesthetic, affective and imaginative—all that springs from the inner resources of the psyche: "the primary, unmodified forces and energies of man, the mysterious springs of Love, and Fear, and Wonder, of Enthusiasm, Poetry, Religion, all which have a truly vital and *infinite* character. . ." This neglected province, the antithesis of the mechanical, Carlyle calls "dynamical." His entire criticism of science rests upon the conviction that we need to develop both of these "great departments of knowledge," and indeed "only in the right coordination of the two, and the vigorous forwarding of *both*, does our true line of action lie." Carlyle continues:

> Undue cultivation of the inward or Dynamical province leads to idle, visionary impracticable courses . . . to Superstition and Fanaticism, with their long train of baleful and well-known evils. Undue cultivation of the outward, again, though less immediately prejudicial, and even for the time productive of many palpable benefits, must, in the long-run, by destroying Moral Force, which is the parent of all other Force, prove not less certainly, and perhaps still more hopelessly, pernicious. This, we take it, is the grand characteristic of our age.[8]

This "grand characteristic" takes the form both of a cognitive and of an institutional imbalance. Within science it manifests itself in a neglect of what once was called Natural Philosophy, and a preference for piecemeal analysis: breaking complex problems down into small, simple, particularized elements, thereby anticipating the tendency of scientific inquiry in our own time in which Gerald Holton recognizes "an asymmetry between analysis and synthesis."[9] Implicit here, too, is a homology between the analytic mode of scientific inquiry and the new principles of economic and social organization. The secular, fragmenting, particularizing tendency within science has its counterpart in the management of the market economy and the new political bureaucracies. All in all, therefore, the mechanical philosophy is producing "a mighty change in our whole manner of existence."

> By our skill in Mechanism, it has come to pass, that in the management of external things we excel all other ages; while in whatever respects the pure moral nature, in true dignity of soul and character, we are perhaps inferior to most civilized ages.[8]

III

Among the misconceptions fostered by C. P. Snow's "two cultures" thesis is the notion that in the twentieth century the humanities have been the province of unqualified hostility to science. My impression is that a comprehensive survey of literary thought, at least, would reveal a spectrum not unlike that which emerged in the age of Emerson and Carlyle. At one extreme, exemplified by the early writings of I. A. Richards, we find what amounts to the emulation of scientific "objectivity" or positivism. In his influential *Principles of Literary Criticism* (1924), and in *Science and Poetry* (1926), Richards embraced a virtual dichotomy between two uses of language, one scientific, the other "emotive." In science, which he conceives as a largely autonomous activity ("the impulses developed in it are modified only by one another, with a view to the greatest possible completeness and systemization, and for the facilitation of further references"), statements are made only for the sake of the reference, true or false, which they allow. This austerely denotative use of language corresponds to that which Emerson and other post-Kantian idealists associated with the Understanding. And like them, Richards opposes it to a connotative or supra-referential use of language for the sake of the effects, both in feeling and in attitude, it occasions. This he calls the "emotive" use of language, a term eloquent in its apparent acceptance of an invidious distinction between the disinterested, "objective" character of scientific statements and the personal bias or "subjectivity" of all other kinds of statements.[10]

It is true of course that Richards later changed his mind and repudiated this early, positivistic phase in his thinking. But it had a long afterlife, particularly within the formalistic "new criticism" which played a leading role in Anglo-American literary thought between, roughly, 1930 and 1960. Whatever their express ideas about the natural sciences (they sometimes were markedly antagonistic), the proponents of this analytic critical method often tended to emulate the posture of the dispassionate, impersonal, scientific observer. Their primary concern was with the "how" as against the "what" or "why" of literature, and in their effort to arrive at precise, neutral, verifiable knowledge, they tended to treat the literary text as comparable, in its susceptibility to precise analysis and in its virtual autonomy, to the isolatable data studied by physicists. Certain academic exponents of the "new criticism" carried the doctrine to extremes never envisaged by theorists like I. A. Richards. They taught students to confine assertions about literary texts to "the words on the page" and to heed the Blakean motto (taken out of context): "To generalize is to be an idiot." (Humanists often seem to associate scientific rigor with an extreme nominalism and avoidance of generalization.) The tacit aim of this kind of literary study moreover, was chiefly to enhance the methodological power of specialists in literary study. Instead of being thought of as a capacity of general culture, available to all educated people, the ability to read imaginative literature was recast by the more extreme practitioners of this new formalism into an arcane skill, like the ability to do physics, accessible only to a tiny minority of expertly trained initiates.

The point about scientism within the humanities—a misplaced application of the assumptions and methods of the natural sciences—is that it can be misleading to gauge the attitudes of humanists toward science by their express opinions alone. It is important to examine what they do (their tacit aims and methods and principles of organization) as well as what they say about science. For they often manage to combine an overt hostility toward the activities of professional scientists, toward science as an institution, with an uncritical and sometimes unconscious emulation of scientific assumptions and procedures. In the realm of literary criticism and scholarship, in any event, the scientistic impulse has remained powerful. After the "new criticism" had lost its vitality, in the 1960s, the yearning of humanists for exact "objective" knowledge, which is to say for a "scientific" critical method, reappeared in such new and ambitious forms as semiotics and a variety of methodological adaptations of "structuralist" principles derived from the latest developments in linguistics.

But the scientistic bent of humanists within the academy was a relatively inconspicuous feature of the cultural history of the recent past. Far more prominent was the new wave of antiscientific thinking that arose in the same period. Following Hiroshima, to repeat, a whole series of problematic science-based innovations had aroused public anxiety about the consequences of scientific discovery. Then the civil crises of the Vietnam era alienated a large segment of the best-educated American youth from any mental work, but especially scientific and technological work, performed in the service of the government or other basic institutions. By the late 1960s, therefore, a large audience was prepared to accept the neo-romantic critique of science at the core of the dissident counterculture.

IV

The viewpoint of Theodore Roszak belongs at the other end of the spectrum of humanist scholars' attitudes toward science from that represented by I. A. Richards and the scientistic literary critics. Taken together, Roszak's two influential books, *The Making of a Counter Culture: Reflections on the Technocratic Society* (1969), and *Where the Wasteland Ends: Politics and Transcendence in Postindustrial Society* (1972),[11] comprise the most systematic effort to formulate a reasoned, coherent ideology expressive of the diffuse antagonism toward science, technology, and scientific rationalism within the dissident "movement" or counterculture which arose during the 1960s.

At first sight Roszak's epistemology would seem to be diametrically opposed to that of those humanists who aspire to exact, "objective" knowledge like that of natural scientists. Yet the striking fact is that a literary theorist like Richards, who in his early writing had endorsed the superior truth value of scientific statements, and Roszak, who regards them as dangerously inadequate, share certain basic assumptions. They both take for granted the antithetical character of objective and subjective, and therefore of scientific and moral (or aesthetic), modes of thought. They both assume that scientific statements are, or come close to being, or provide a compelling illusion of being, "objective." Or, to put it even more subtly, if total objectivity is not finally attainable, Roszak asserts, the fact remains that scientists can still feel and behave as if it were. If, he says, "an epistemology of total objectivity is unattainable, a *psychology* of objectivity is not. There is a way to *feel* and *behave* objectively, even if one cannot *know* objectively."[12]

The apparent objectivity of scientific knowledge is crucial, it would seem, to both the emulation and the antagonism it elicits from humanists. Whereas Richards (in his positivist phase) implied that objectivity conferred a superior authority upon statements made by scientists, Roszak believes just the opposite. He concedes that in the eyes of the gullible public scientific knowledge has immense authority, but it is a misplaced authority. To insist upon the quantifiability and verifiability of knowledge is, on this view, to insure its shallowness and triviality. The experimentally verifiable results of scientific inquiry comprise a body of information of undeniable utility for the mastery and manipulation of the biophysical world. But when such information is deferred to as the exemplar of true knowledge the results can be disastrous. For the "scientific picture of nature" it provides effectively screens out all qualities of mind and nature, all modes of perception and of being, except those with instrumental value.

> Objective knowing gives a new assembly line system of knowledge, one which relieves us of the necessity to integrate what we study into a moral or metaphysical context which will contribute existential value. We need no longer waste valuable research time and energy seeking for wisdom or depth, since these are qualities of the person. We are free to become specialists.[13]

The scientific style of mind, devoted as it is to "objective knowing," is the radical flaw in the culture of urban industrialism. Roszak calls this style, after William Blake, "the single vision": a one-dimensional, technologically useful, but humanly impoverished world view. Although it happens to be the one form of experience now dignifiable as knowledge, it should not in his view be called that. It would be more accurate, Roszak contends, to call what science reveals to us about nature "information," and to reserve the term "knowledge" for those holistic, often ecstatic syntheses of fact and value—of nature, spirit, and self—which are properly called "gnosis."[14] His epistemology, therefore, must be distinguished from that held by the moderate romantics (Wordsworth, Coleridge, Emerson) who envisaged an accommodation between the two modes of perception, empirical and transcendental, which could effect a "marriage," in Emerson's figure, between matter and mind. Again, Roszak's version of the neo-romantic critique of science is like the positivism of a Richards in seeming to rule out the potential complementarity between the two kinds of knowing.

Although Roszak draws upon the romantic poets, especially Blake, and indeed upon the entire legacy of visionary and prophetic literature going back to the Old Testament, he presses the case against practical reason to a new extreme. Many other writers have insisted upon the superiority of intuitive, nonrational ways of knowing; many have pointed out the severe limitations of scientific rationalism; but few before now have singled out the scientific world view as *the* root cause of what is most alarming about modern societies. According to Roszak, however, it is the critical variable in an essentially destructive, perhaps suicidal, pattern of collective behavior.

> Undeniably, those who defend rationality speak for a valuable human quality. But they often seem not to realize that Reason as they honor it is the god-word of a specific and highly impassioned ideology handed down to us from our ancestors of the Enlightenment as part of a total cultural and political program. Tied to that ideology is an aggressive dedication to the urban-industrialization of the world and to the scientist's universe as the only sane reality. And tied to the global urban industrialism is an unavoidable technocratic elitism.[13]

The notion that rationality, or the quasireligious belief in Reason, is the motive force behind urban industrialism exemplifies Roszak's idealistic theory of history. Unlike most contemporary historians, he imputes to ideas an almost exclusive efficacy in social change. He therefore portrays the contemporary world as the scene of an all-encompassing Manichean struggle between opposed views of reality, each marked by an ideal type of knowing: scientific rationalism and gnosis. One is reductive, partial, analytic; the other augmentative, holistic, synthetic. The social forms accompanying each are, for his purposes, largely irrelevant. So far from being important determinants of human behavior, indeed, social structures and processes are for Roszak relatively inconsequential reflections of the dominant mental style. What chiefly accounts for differences in ways of life, accordingly, are differences in the ruling conceptions of reality. Roszak's theory of history might be called a form of metaphysical or, to be more specific, epistemological determinism.

Assuming that theories of knowledge are the prime movers of history, Roszak deals with them apart from the social groups which embraced them or the functions they served in actual historical situations. This enables him to discuss the ecstatic, visionary mode of cognition (gnosis) "found in the world's primitive and pagan societies" without reference to its uses as an instrument of minority rule or, in many instances, of tyranny. The political role of the shaman is, so far as he is concerned, largely irrelevant; what matters is the shaman's "unitary vision bringing together art, religion, science, and technics."[15] Similarly, he discusses the emergence of the purposive-rational way of knowing in the Enlightenment without any reference to the larger vision of scientific progress as a corollary of political, social, and psychological liberation. My impression is that Roszak's apolitical sense of history as a battlefield of free-floating ideas is characteristic of the view held by many adherents of today's counterculture.

The inadequacy of this simple, single-factor mode of historical explanation is nowhere more apparent than in Roszak's attempt to account for the destructive uses to which our society puts scientific knowledge. Whereas Whitehead's description of the romantic protest against science had allowed for the distinction between science as a mode of cognition and science as it functions in a particular social setting, Roszak's protest does not. On the contrary, his fundamental charge against a science grounded in instrumental reason is that the evil uses to which it is put follow from its epistemological inadequacy. But it is not clear whether he considers those evils a necessary or merely a possible consequence. Although his generalizations imply that scientific knowledge leads inevitably to flagrant abuses of mankind and of nature, his specific examples are ambiguous.

> We should by now be well aware of the price we pay for regarding aesthetic quality as arbitrary and purely subjective rather than as a real property of the object. Such a view opens the way to that brutishness which feels licensed to devastate the environment on the grounds that beauty is only "a matter of taste." And since one person's taste is as good as another's, who is to say—as a matter of *fact*—that the hard cash of a strip mine counts for less than the grandeur of an untouched mountain?[16]

What does it mean to say that rationality "opens the way" for strip mining? Roszak's point is that the sharp instrumental focus of modern science ignores the aesthetic attributes of the object. To protect the environment, to give mountains and trees adequate "standing" in our culture, we need to restore a sense of the absolute value inherent in natural objects comparable to the divinity imputed to them by primitive (animistic) modes of thought. But Roszak does not give us much help in imagining a mode of knowledge capable of coordinating a geological understanding of a mountain with an apprehension of its allegedly inherent beauty. We have reason to believe, for one thing, that the beauty is not inherent. As Marjorie Hope Nicolson demonstrated years ago, most English travelers before the late sixteenth century regarded the Alps as ugly excrescences on the face of nature.[17] Had they been practitioners of gnosis they presumably would have advocated strip mining on the slopes of Mont Blanc. In one sense, admittedly, the example is absurd. Roszak's point is that gnosis, by definition, entails a world view incompatible with either modern geology or strip mining. Yet the absurdity does serve to illustrate the all-or-nothing character of the choice we are being invited to make. No accommodation between science as we know it and Roszak's conception of an adequate epistemology is conceivable.

To say that rationality "opens the way" for strip mining is in any case far from saying that the resulting devastation is attributable to science. Let us suppose that advances in geology are among the factors that have contributed directly to the feasibility of strip mining. It is still necessary to consider the relative influence of geological knowledge and economic profitability (Roszak's "hard cash") as motive forces here. Granted that technical competence (equated with "rationality" hence "science" in this lexicon) makes strip mining possible, the fact remains that a business corporation conceives and organizes the operation and a juridical system legitimizes it. In what sense, therefore, is science accountable here?

Roszak's answer embodies the crux of the countercultural critique of science. It is worth noting, incidentally, that he did not address the question in the original draft of this passage. Following the conference in which it was criticized, however, he added this telling afterthought:

> Is such barbarism [i.e., strip mining] to be 'blamed' on science? Obviously not in any direct way. But it is deeply rooted in a scientized reality principle that treats quantities as objective knowledge and qualities as a matter of subjective preference.[18]

In other words, the technique of the mining engineer and the economic calculations of the corporate management, like the scientific information of the geologist, may be thought of as products of the same "scientized reality principle." An old-fashioned historian's distinction between the enabling power of scientific knowledge and the motives generated and sanctioned by socioeconomic institutions is not meaningful to Roszak. According to his idealistic interpretation of history, *all* of these activities are traceable to the one root cause: a rationalistic world view. Since the domination of that ideology is a result of the advances of scientific knowledge, Roszak is in fact putting the ultimate blame for the destructiveness of modern society upon science.

To sum up, then, the strip miner's brutish devastation of the landscape typifies this conception of the way our flawed metaphysic issues in social evil. The epistemological flaw, again, is the reductionism characteristic of science: the screening out of those qualities of mind and nature, in this case aesthetic qualities, not useful to the purpose at hand. If one accepts the major premise of Roszak's metaphysical determinism, his apocalytic conclusion follows logically enough. The scientific view of man's relations to nature is conducive to a kind of institutionalized moral nihilism. Hence the destructiveness of urban industrial society is irremediable, and our only hope is to replace it and the conception of reality from which it derives.

V

Today's criticism of science has a long history in Western thought. To be sure, a series of shocking events following World War II aroused widespread public anxiety about the latest advances in research. In one sense, therefore, the counterculture's attack upon rationality may be interpreted as an extreme expression of a current mood. But it is necessary to recognize that this recent development also is a new phase of the "romantic reaction" that began some two centuries ago. Then, as now, the reliability of scientific knowledge within its own proper sphere was not in question. Many thinkers noted, however, that the scientific view of reality imputes excessive importance to the small part of life susceptible to experimental and logical methods of analysis. Even a writer like Emerson, who retained much of his Enlightenment faith in scientific progress, expressed a characteristic post-Kantian skepticism about the sufficiency of practical reason. He did not doubt the absolute validity of "natural facts," and since he regarded "Nature" as "the present expositor of the divine mind,"[19] he believed that knowledge grounded in empirical facticity could in theory yield the kind of certainty and authority hitherto claimed for religious truth. But in order to satisfy the full range of human needs, it would be necessary to "marry" the neutral data to value-laden concepts arrived at by the other (intuitive, mythopoeic, holistic) way of knowing.

By thus insisting upon a coordination of the two modes of cognition, even the more optimistic critics of empiricism helped to prepare the way for today's neo-romantic attack upon science. For the anticipated marriage of fact and value, matter and mind, did not occur. So far from effecting a closer, more meaningful and harmonious relationship between man and nature, science in the context of Western industrialism is perceived by its latter-day critics as having divested nature of its ultimate, or teleological, significance. At bottom, then, this critique of science is rooted in an essentially religious, suprasensual, conception of man's relations with nature. We must recover our capacity for gnosis, according to Roszak, because we desperately require access to the value, meaning, and purpose presumed to reside in "things as a whole."[20] The monster created by science is meaninglessness. Instead of providing the unconditioned meaning sought by mankind, science in the nineteenth century came to be identified with new, more acute forms of alienation from nature.

The dislocation attendant upon the rapidly accelerating rate of industrialization and urbanization was destined, in the long run, to undermine confidence in science. Whether science "caused" these changes, or whether specific technological innovations did in fact derive from scientific research, is largely beside the point. Scientists, inventors, and entrepreneurs, as they functioned within an expanding capitalist society, would appear to critics of that society as kindred embodiments of the same predominantly analytic, secular, matter-of-fact mentality. The emergence of entire industries manifestly based upon recently acquired scientific knowledge, the electrical and chemical industries in particular, subtly eroded the ideal of science as the disinterested pursuit of truth. Idolators of "progress" boasted about the complicity of scientists in changing the face of nature. By the end of the nineteenth century the partnership of science, engineering, and capitalism was acknowledged by its defenders and critics alike. The old distinction between pure and applied science had lost much of its force long before the neo-romantic critique of science had been formulated.

These developments also were to lend more and more credence, as time went on, to Carlyle's choice of the machine as the cardinal metaphor for the emergent industrial system. Scientific knowledge, according to that figure, is the intellectual fuel upon which an expanding machinelike society runs. In the iconography of antiindustrialism, therefore, machines represent the most conspicuous products, both physical and cultural, of modern science. They simultaneously represent several aspects of science-based technology: (1) a new kind of apparatus and technique; (2) the rational organization of work and of economic activity generally; (3) the principles underlying the first two senses of "technology," which is to say the analytic mode of thought and, by extension, the social order typified by a perfection of means and a diminished control over ends.

Beginning with Carlyle's generation, novelists and poets invoked the imagery of mechanization to convey a sense of dwindling human agency, or what may be called, in retrospect, the incipient totalitarianism of industrial society. By the 1960s adherents of the dissident movement throughout the West were invoking the terms "technology" and "the system" and "the machine" more or less interchangeably. All referred to the controlling network of large-scale institutions (government, business corporations, universities) in whose services most scientists do their work. The "machine," in other words, is coterminous with organized society. We are reminded of the inception of the Berkeley uprising in 1964 when one of the leaders, after describing "the operation of the machine" as intolerably odious, called upon his fellow students at the University of California to throw their bodies upon it, if necessary, to make it stop.[21] What is striking about the episode is the extent to which the audience, and the rebellious youth movement in general, seems to have accepted the meaning tacitly imparted to "The Machine."

The received iconographical convention by which "the machine" of science-based high technology is equated with organized society bears witness to the deep-rooted, historical basis for the antiscientific strain in contemporary culture. As Whitehead noted years ago, the literary reaction to the scientific revolution of the eighteenth century expressed the "deep intuitions of mankind penetrating into what is universal in concrete fact." He insisted, moreover, upon the philosophic cogency of the great body of imaginative literature which testifies to the discord between those intuitions—aesthetic, moral, metaphysical—and the "mechanism of science."[22] The discord reflects an increasingly obvious discrepancy between what science provides in the way of certain, verifiable knowledge, and what mankind would have in the way of a meaningful existence. In view of the history of the half-century since Whitehead made these observations, and of the ambiguous part that science has played in that history, it is not surprising that the discord has grown sharper. Nor is there any reason to doubt that it will continue to do so. It would be a serious mistake, accordingly, for those concerned about the future of science to underestimate the appeal, or the force, of the neo-romantic critique of science as a mode of knowing built upon an inadequate metaphysical foundation.

REFERENCES

[1]"Some Thought on the Other Side of This Life," *The New York Times* (April 12, 1973): 45.

[2]In 1974, a survey commissioned by the National Science Foundation showed that 39 percent of the American people expressing any opinion do not agree with the proposition that "Overall, science and technology do more good than harm." See Loren Graham, "Concerns about Science and Attempts to Regulate Inquiry," in this issue of *Daedalus*.

[3]Although the writers I have in mind seldom formulated the distinction between the two referents of the word "science"—science as a mode of perception or inquiry, a way of knowing the world, and science as organized activity, a way of behaving in society—it is often implicit in their thought. By now, in any case, its importance cannot be overstated; my impression is that much of the confusion surrounding discussions of the "limitations of science" derives from the tendency to conflate these two meanings of "science."

[4]*Science and the Modern World* (New York: Macmillan, 1947), p. 138.

[5]M. H. Abrams (ed.), *Norton Anthology of English Literature* Vol. 2 (New York: Norton, 1962), p. 89.

[6]*Complete Works*, Vol. 1 of 12 vols. (Boston: Houghton, Mifflin, 1884), p. 42.

[7]"Science and the Sense of Self: The Medieval Background of a Modern Confrontation," in this issue of *Daedalus*.

[8]"Signs of the Times," *Critical and Miscellaneous Essays* (Chicago: Bedford, Clarke, n.d.), p. 21.

[9]"Analysis and Synthesis as Methodological Themata," *Methodology and Science* 10 (March, 1977): 3–33.

[10]See especially chapter 34, "The Two Uses of Language," *Principles of Literary Criticism* (New York: Harcourt Brace, 1950), pp. 261–271.

[11]*The Making of a Counter Culture: Reflections on the Technocratic Society* (New York: Doubleday, 1969); *Where the Wasteland Ends: Politics and Transcendence in Postindustrial Society* (Garden City, N.Y.: Doubleday, 1972).

[12]*Ibid*, p. 167.

[13]*Ibid*., p. 171.

[14]"The Monster and the Titan: Science, Knowledge, and Gnosis," *Daedalus* (Summer 1974): 17–32.

[15]*Ibid*, p. 27.

[16]*Ibid*, p. 25–26.

[17]*Mountain Gloom and Mountain Glory* (Ithaca, N.Y.: Cornell University Press, 1959).

[18]"The Monster and the Titan," *op. cit.*, p. 26. This passage did not appear in the first mimeographed draft of the paper which was discussed by the present writer at a conference sponsored by *Daedalus*.

[19]*Complete Works, op. cit.*, p. 68.

[20]"The Monster and the Titan," *op. cit.*, p. 21.

[21]Seymour Martin Lipset and Sheldon Wolin (eds.), *The Berkeley Student Revolt*, (New York: Doubleday Anchor, 1965), p. 163. For the background of this iconographical tradition, in American and English literature respectively, see Leo Marx, *The Machine in the Garden: Technology and the Pastoral Ideal in America* (New York: Oxford, 1964) and Herbert Sussman, *Victorians and the Machine* (Cambridge, Mass.: Harvard University Press, 1968).

[22]*Science and the Modern World*, *op. cit.*, pp. 126–127.

DON K. PRICE

Endless Frontier or Bureaucratic Morass?

ENTANGLED IN THE PROCEDURAL CONSTRAINTS that go with government money, some scientists are almost tempted to regret the implicit bargain on which massive federal research support has been based since the Second World War. The political strategy they have followed since Vannevar Bush's classic report of 1945, *Science the Endless Frontier*,[1] may need to be thought through again in the light of new circumstances and new public attitudes.

It may be that the world of science is not unique in this way, but only a few jumps ahead of other institutions in American society. C. P. Snow, we may recall, argued that scientists had the future in their bones. It may unfortunately, in matters of politics, be more in their bones than in their brains, which often contain the same obsolete political myths in which other Americans devoutly believe.

But there is no doubt that the scientific community is ahead of the pack in one respect, and one which has forced it to rethink its relationship to politics. While business leaders speculate about the future limits to economic growth, American scientists have already had to adjust to a static or even a declining level of financial support.

In the scholarly world, it is fashionable here and there to argue that it will be a good thing to disavow materialistic values, at least for the world of business and industry, and to accept with equanimity a static GNP, declining use of energy, and the philosophy of "small is beautiful." It is somewhat more rare to find scientists who apply the same line of argument to the financial support of research. On the contrary, the scientific community generally is aware that the reduction in federal support, and the resulting cessation of its economic growth, make it much more difficult to deal with many problems in the administration of research programs, and indeed in the academic world as a whole.

Perhaps the most troublesome of these problems is the recent increase in legal and administrative constraints that severely limit the autonomy of university administration and the freedom of research workers—constraints that are imposed not in the interest of the quality of research but of short-term practical results, regional distribution of funds, or other criteria more or less irrelevant to scientific excellence. Almost equally annoying to scientists is the change in political attitudes; members of Congress now tend to look on them as just another selfish pressure group, and not as the wizards of perpetual progress.

In general terms, it would be easy to get agreement within the scientific community on the remedy: more generous federal support on a longer-term basis with fewer irrelevant conditions and regulations, and more recognition of the principle that the excellence of research in American universities is a matter of profound national interest.[2] But agreement in general terms by institutional executives does not provide effective political support when their constituency is apathetic or divided.

University research scientists (this paper will concentrate on university research, and neglect the important world of industry, which in recent years has been cutting back on support of basic research) seem in recent years to have taken for granted a degree of freedom and financial support that depended on a novel and fragile arrangement. It has been easy to forget the atmosphere of the time, less than a half-century ago, when the federal government considered research grants to private universities improper if not unconstitutional, and when the leaders of the National Academy of Sciences objected on principle to letting private universities accept government funds.[3]

Many scientists who have become addicted to federal support are now inclined to wonder whether their old leaders were right. While they are unable to swear off periodic injections of government grants, they repent of their addiction, and blame the bureaucratic constraints they must endure on the same three bogies that frightened scientists in the 1930s. These, in caricature, are politicians (by definition bad people), antiintellectuals who wish to impose ideological or theological constraints on research, and a bureaucracy dominated by an overmighty executive.

It would be more plausible, it seems to me, to blame undue constraints on causes that are almost the opposite in nature. This list, equally a caricature and not to be defended literally, would put the main causes thus: first, a political strategy for the support of science that was devised by scientists themselves and based on the experience of private philanthropy; second, the acceptance by political authorities of science as the dominant intellectual approach to public issues, which scientists and other liberal intellectuals agree must therefore be regulated in the public interest; and third, a constitutional structure too decentralized to sustain the integrated and long-term view of public policy which might justify the support of science as an intellectual and educational enterprise.

I would be more inclined to defend this second list of causes than the first, but both have some elements of truth, and neither is likely to serve as the basis of a constructive political strategy, which requires something better than mutual recrimination between the worlds of politics and of science.

As a first step toward thinking through such a strategy, let us recall the way in which the present research support system developed after World War II.

Politics as the Source of the Problem

Politics is often thought of, in derogatory terms, as nothing but the way in which people get power through the process of popular election. That kind of politics had remarkably little to do with the change from 1940, when the federal government paid for one-fifth of the nation's research, to 1975, when it paid for well over half. The growth in support for science was never a big electoral or partisan issue.

But there are other bases for political power. Religion and wealth are among the older ones. And the experience of the past generation or two teaches us that newer forms of social organization are politically powerful, notably the professions and the scholarly and academic institutions. Organized knowledge as a political force should no longer be underestimated.[4]

The conversion of the United States federal government to the support of scientific research after World War II was a brilliant feat of political strategy. Its quality should not be measured in terms of dollars alone. Other countries greatly increased their financial support of research over the same period. But nowhere else did a group of scientists manage to persuade their governments and their research institutions not only to give up their scruples against public support of private science, but also to work out the arrangements for distributing the support that maximized the political influence of those whose authority was based on knowledge rather than on votes. These arrangements depended on a tacit bargain that dealt with both political ends and means.

As for the ends, the Congress had to be persuaded that basic research was a necessary foundation for practical progress, and that applied results would flow automatically from it. With the wartime experience as evidence, this argument, the core of the Bush report, proved convincing. It led to a continuing bargain by which scientists would get support for basic research which government officials would hope would lead to applied research and on to useful developments.

As for the means, it seemed necessary to free basic science in several ways from the constraints inherent in the older federal research programs, such as agricultural research in the state experiment stations and research in government laboratories like those of the National Bureau of Standards. The main constraint, of course, had been the exclusion of private universities from the use of federal funds. But if they were to be included, it was important to give them more freedom than could be provided under the agricultural research pattern, with its emphasis on the distribution of funds among the states by statutory formulas that gave no weight to differences in quality,[5] or than could be provided for the government laboratories, which got their money through line item budgets by grace of congressional committees. Both arrangements led to heavy emphasis on applied rather than basic research.

To get away from such constraints, a three-part tactic was adopted. It was not, I suppose, expressly formulated as a tactic by anyone. But all three of its parts reflected the experience of the scientific leadership of the time with private foundations,[6] which had proved the most effective way of organizing financial power so as to give weight to scientific values. It must have seemed only natural to use similar arrangements to give weight to organized science as a counterbalance to the power flowing from the electorate. What worked with Rockefeller and Carnegie money ought to work with congressional appropriations.

The first tactic was to combine research with university teaching. The United States was not unique after 1945 in increasing support for research and development, but it was unique in the extent to which it supported such work through grants to universities. The separate research institute dominates basic research in the Soviet Union and many other countries; even in the United

Kingdom and Canada the universities get a smaller share than in the United States of the government's total research program (including applied research).

The tactic had its precedents in private philanthropy. Just after the First World War, the National Research Council's negotiations with the Rockefeller Foundation for the support of science turned initially on hopes for a series of special research institutes in physics and chemistry. There were, after all, the models of the Carnegie Institution of Washington and the Rockefeller Institute for Medical Research. But the Rockefeller Foundation and the National Research Council threw their weight behind the NRC fellowship program as the best way of strengthening basic science in the one setting most free of commercial self-interest or political pressure—the university.[7]

The university seemed then, and still seems, the most stable basis on which to defend the independence of science. With independent endowment funds and a senior faculty on permanent tenure, it was in a position to defend freedom of inquiry against popular passions or economic self-interest. State universities, more vulnerable in some ways because of their financial dependence on legislatures, came to imitate the private universities in the security afforded their permanent tenure faculty. (Much later, in 1955, even when dealing with so political an issue as military strategy, it seemed reasonable to try to give the new Institute for Defense Analyses a status independent of the source of its funds by establishing it under a consortium of universities.)

The second tactic was the project grant. It had been developed in various private foundations, most notably the Rockefeller Foundation and the Carnegie Corporation, which shortly began to lose interest in making endowment grants. The main purpose of the project grant was doubtless to make it possible for a foundation with specific purposes to make sure that its funds were devoted to them, and not merged with more general programs, or abused by faculty politics, or neglected by weak university administrations. To the scrupulous foundation executive, the record of universities in using general grants seemed discouraging. The project grant seemed a useful and efficient way to influence the conduct of research in the direction of social purposes, and away from excessive disciplinary specialization.

But a procedural effect was to give more influence to the professional staff of a foundation by comparison with its donor or trustees. In order to make an institutional endowment, a single trusted advisor to a great philanthropist, or at most a tiny staff, could negotiate a major gift and a philanthropist or his trustees could be expected to understand it and make the essential decision. With a project grant dealing with a highly specialized science, most trustees had to leave the judgment to their expert staffs, and have staffs large enough to follow up on the project, or even play some role in its direction. Some of the older generation of foundation executives were contemptuous of this trend, believing that the better role for philanthropy was to build independent institutions and leave the specialized decisions to them,[8] but in recent years the project grant's apparent advantages led the major new foundations, such as the Ford and the Robert Wood Johnson foundations, to adopt it as their main mode of operation.

For the federal government just after World War II, the project grant seemed appealing to scientists because it offered a tactic to avoid detailed congressional control of funds, such as prevented the Department of Agriculture or the Bureau of Standards from becoming preeminent centers of basic research. Just as in private philanthropy, it would be hard to persuade those with ultimate power (in this case, members of Congress) to defer to experts in deciding on broad program grants or institutional grants. But if the grants were to support highly specific and specialized work, Congress could be persuaded that the merit of the project and of the investigator were the only appropriate criteria, and that such merits could be judged only by scientific peers. Moreover, it let the Congress sidestep various troublesome issues: through the project grant, support could be given to universities without adopting a general program of aid to higher education or discriminating between state and private institutions, and above all without facing the issue of church and state. But incidentally, it put the staff members of the granting agencies (themselves under pressure from congressional committees) in a position to review the conduct of research projects. And it probably weakened the influence of university administrations and the commitment of individual scientists to their institutions, by comparison with their commitment to the sources of their research grants.

The third tactic was the pattern of organization, including the personnel system, by which the community of science was to be given a political authority that did not depend on popular votes. The most obvious part of this tactic was the creation of the National Science Foundation and its board, on the model of the private foundation, and the vesting of substantial authority by law in the advisory councils of the National Institutes of Health.[9]

But the most striking shift in political influence to the scientific community did not come from the board or committee pattern of organization. It came instead from the system of personnel, and from the support of the processes of policy planning by part-time advisers under government grants and contracts. Before World War II Congress was still prejudiced against letting anyone other than full-time government officials gain any control over governmental decisions. This prejudice was breaking down slowly. Civilian departments, which before the New Deal had typically no statutory authority to employ part-time consultants, had been lured into using them by the offer of philanthropic funds for the purpose, and had continued the habit with subsequent congressional sanction.[10] But the great increase in the use of consultants and advisory committees came after the Second World War, as those scientists who had served in the war agencies went back to their private status, but continued to maintain substantial influence over government programs in the vast pyramid of committees and advisory councils of which the President's Science Advisory Committee became the apex. Working from a tenured position in a university, a scientist could, on a part-time basis, play a more painless and perhaps a more influential role in public policy by membership on advisory committees, by participation in summer study projects or formal "evaluations" of current programs, and by serving as an individual consultant to short-term and harried political executives, than if he accepted the burdens and responsibilities and vulnerable status of a full-time position.

These three tactics together made a major shift in the balance of political authority between the type of politics based on votes and the type of politics based on knowledge or scientific status. But the authority gained by the scientific community was not of the kind that could be defended as a matter of constitutional right; it was a delegated authority, and it depended on the continued confidence among elected politicians in the assumption on which the tacit bargain was founded—that basic research would lead automatically to fruitful developments.

It is obvious that that confidence was shaken during the 1960s, and not only among members of Congress. The change in congressional attitudes may have been affected somewhat by the general malaise in the intellectual world, some parts of which, including scientists themselves, were beginning to doubt the omnicompetence of science in human affairs. But it probably resulted more from the normal disposition of anyone who lends or grants money to want to know what use is being made of it, and whether the terms of the bargain are being kept. As universities became very big businesses indeed and dependent on research grants for their budgets, and as professors became simultaneously advisers on government policy and salesmen for their project applications, each of the three tactics that had given the scientific community such status just after the Second World War came to be looked on with cynical suspicion by government administrative officers and members of Congress alike. One may speculate whether the general status of basic science might not have been protected better if more professors had concentrated their political energies on advocating the case for general support with fewer restrictions, rather than on advocating particular issues of public policy or soliciting specific project grants.

The problem is of course a political problem, in the general sense of the word "political." But it would be a mistake to think that it would be solved if scientists never took another grant from the government. The patterns of governmental organization and procedures for the support of basic research were installed in government by the political action of scientists, in imitation of the patterns of private philanthropy. As government agencies have tended in recent years to put more conditions on their research grants, so have private foundations. The most restrictive controls are subtle and not reduced to contractual terms (just as the congressional restrictions on executive bureaus are often in the legislative history rather than in the provisions of statutes). Accordingly, I think it would be difficult to measure the degrees of constraint that go with government research grants by comparison with those from private foundations or corporations, but from my own experience I would guess that the private grants are on the average more restrictive; private foundation staffs may have more prestige, and certainly are granted more leeway by their own superiors, than their governmental counterparts. And they are paid more.

The comparison is not very relevant, since the volume of government support is far beyond the capacity of private institutions to sustain. Private foundations have long since unloaded on government their most expensive areas of research support. But if such support could conceivably be transferred back to foundations and business corporations, one must remember that what they too do in philanthropy is now controlled, in some degree, by government as a matter of tax policy, administrative regulations, and contractual conditions. Government agencies now behave more like private institutions than they did before the Second World War, and private institutions are now more like government agencies in their subordination to public policy. Science and technology, by demonstrating the flexible possibilities of administering government programs through grants or contracts to private institutions (remember the atomic bomb and the man on the moon, as does every member of Congress), have had a lot to do with the change, which scientists are less inclined to appreciate than they might be.[11]

The Purposes of Constraints on Research

In the rich variety of American politics, there are plenty of examples of benighted and reactionary politicians who distrust intellectuals and academic institutions. Accordingly, it is a great temptation to blame them for any or all of the constraints which public authorities have placed on university science in recent years. But on the whole it seems likely that the constraints about which universities and scientists complain the most are the result of liberal reforms and egalitarian pressures, which are advocated more generally by intellectual reformers than by backwoods reactionaries. Let us look at the apparent purposes of the several types of constraints.

REGULATORY RED TAPE

The constraints that have been most burdensome in financial and administrative ways have had little to do with, or at least were not intended to affect, the substance of scientific work. They have included the clerical and managerial burdens of social security taxes, equal opportunity and affirmative action requirements, environmental protection, occupational health and safety, and free access to information (such as the proceedings of advisory committees or letters of recommendation) previously considered confidential. These were originally considered liberal reforms. As they come to impinge on universities as well as on industry, it is interesting to observe how in liberal university circles there is coming to be more tolerance for the belief in laissez-faire policies that was formerly considered the mark of benighted reactionaries.

It is hard to imagine that research, whether in universities or in industry, can be completely exempt from regulations that government imposes on other institutions in what has come to be considered the public interest. But it is possible to consider ways in which such regulations may be adjusted to the peculiar circumstances of scientific research or of academic administration. Before the Office of Scientific Research and Development undertook the support of wartime research in universities by the contract device, no one assumed that government research contracts could be let in broad terms, omitting the detailed procedural checks that material procurement contracts contain to protect the taxpayer. But such contracts were devised and used, some of them giving universities as much freedom as grants. As time went on, however, and as competition for funds increased and abuses in university management were discovered by zealous government investigators, the terms of contracts were tightened by various agencies.[12]

It should be possible for government agencies to loosen up the red tape that now entangles the grant and contract procedure, on grounds that the nature of research work is such that detailed requirements are counterproductive and cost the government more in terms of final product than they are worth. But contracting officers themselves are under pressure from congressional committees to prevent misuse of federal funds, and their best protection is often to apply uniform rules, or to base requirements on objective and quantitative criteria. Both tendencies make enforcement easier, and preclude legalistic objections. But while eliminating arbitary authority, they also eliminate the use of good judgment or the consideration of general values that may be significant but hard to quantify. "Zero tolerance" of carcinogens, or uniform statistical goals for affirmative action, are good examples.

It is unlikely that scientists can be totally exempt from the constraints that regulatory red tape imposes on all other parts of society. Their relief may come in part as their institutions prove that they are capable of policing their own performance more adequately, and in part as the government improves the general quality of its regulatory activities, giving more weight to broad judgment and less to detailed controls. The best example of such improvement in recent years has been in an agency that oversees other government agencies: the General Accounting Office, which in 1946 had a staff of 14,904, most of them clerks and auditors and only three percent with professional qualifications, had by 1976 cut its total staff to 5,351 by raising their quality (seventy-two percent were by now professionals) and delegating the detailed checking of accounts to the agencies that it supervises.[13]

If a similar type of improvement can be made in the supervision of grants and contracts with universities and other private institutions, which are equally entitled with federal bureaus to argue that they can spend government funds more efficiently under broad delegation rather than detailed review, it can be done only if the public officials concerned can be less numerous, of higher professional status and quality, and more secure politically in their ability to make discretionary decisions, rather than having to justify their actions by narrow and legalistic procedures.

THE ATTACK ON ELITISM

A more general constraint on the quality of university research comes from the attack on elitism in the distribution of research funds. The project grant system, and the "peer review" method by which it distributed funds, produced a concentration of grants in a limited number of universities. The tactic of the project system had worked; by contrast with the agricultural research program, money had been distributed not by geographical formula but by judgment regarding excellence through "peer review," in the original meaning of "peer." In an American jury trial, the judgment of one's peers means judgment by a random selection of citizens; but scientific "peer review" goes back to the original meaning, in which one should be judged only by those with competence to pass judgment—one's equals.

In reaction during the 1960s against the apparent concentration of federal grants, the National Science Foundation, the National Institutes of Health, and other grant-making agencies devised procedures to distribute funds more equally, such as making institutional or broad departmental grants, or giving funds for advanced fellowship training to institutions rather than directly to individuals who would tend to congregate in a small number of universities with highest prestige. During the same period, various states undertook to shift their expenditures for higher education away from the major research universities, and into the support of community colleges, while federal expenditures for postsecondary education shifted toward remedial and vocational education.[14]

These shifts in policy seemed to work to the disadvantage of the leading research universities. Excellence in fundamental research is not something that can be distributed by statute uniformly among the states and cities. In some fields of knowledge, private foundations had discovered that some subjects could be developed only by a very limited number of universities, and that the effort to spread the work would prevent the development of enough "critical mass" to be effective; certain types of foreign area studies are good examples. If there are about two hundred universities that grant the Ph.D. degree, that is probably more than can be justified on a national perspective. But should the number of major centers, receiving federal support, be ten or be fifty or a hundred? There is probably no good way to answer that question with confidence, but two observations about the process of considering it may be pertinent.

First, professional politicians are not the original source of the egalitarian or antielitist movement. In such matters, members of Congress reflect opinions more than they invent them. The community of science is not homogeneous. Opinions as to the merit of concentration of research resources are not the same at a meeting of the National Academy of Sciences as at the American Association for the Advancement of Science, the one representing the research peerage, the other open to anyone with a twenty-eight dollar membership fee and thus constituting, as Gerald Holton has remarked, the lower house of the parliament of science. And they clearly differ between the Association of American Universities and the National Association of State Universities and Land-Grant Colleges. It would have been a miracle if the concentrating effects of the political tactic of the project grant had not been noted with envy by colleges and universities which were coming out badly in the competitive system, and if they, constituting a considerable majority of institutions, had not been able to muster political pressure in favor of a more egalitarian distribution of funds.

Second, no one knows how to make and defend a frankly elitist decision about the distribution of federal funds to the best universities, within the American constitutional system. To make such a decision requires the existence of a central institution with great discretionary authority, or with a settled national tradition to guide it. In Great Britain, the University Grants Committee has made such decisions, but it was set up so that it was shielded from questioning either by the House of Commons or the Treasury; to imitate its effects, the United States had to adopt an apparently opposite pattern through the project grant system. In California, the state universities get preferential treatment over the state colleges and the community colleges, but the universities themselves are distributed regionally so as to reduce geographical jealousies.

The heads of the major research universities themselves are more aware than anyone else of this point; the recent plea of fifteen of them for more federal research support concluded that is was impossible for it to be expended through institutional grants to "the jostling and ever-changing set of universities that someone thinks are at the top."[15]

CONTROL OF SUBSTANCE OF RESEARCH

Much more dangerous to the essential freedom of science than regulatory red tape, or than pressure for the egalitarian distribution of funds, is the kind of political pressure that threatens to bring about thought control or ideological conformity, by constraints on the substance of research.

Some such constraints arise in ways that harass individuals but are only minor irritants to the scientific community as whole, such as the publicity tactic of a legislator who ridicules, to impress his constituents, the title of a research project that seems absurd or overly technical to the lay reader. Others deal with issues of public health, safety, or morals, as do regulations dealing with DNA experiments, or medical experiments on prisoners, or experiments involving cruelty to animals. These issues may be troublesome and difficult, but generally they do not involve political efforts to control the outcome of research or its inner methodology, or to impose on it a political ideology. And they are stirred up as often by scientists themselves as by politicians, as in the case of DNA.[16]

The greatest challenge to the use of scientific methods has come in attacks on the social sciences, which seem to the layman more obviously to threaten some types of political or religious values. A generation ago, this threat seemed ominous enough. The congressional investigation headed by Congressman B. Carroll Reece from Tennessee undertook to prove that private foundations ought to lose their tax exemption because they supported the social sciences, which were promoting socialism within the United States.[17] These seem, in retrospect, more a farce than a danger, but they were taken seriously enough by social scientists at the time. That was the period in which the National Science Foundation, perhaps recalling the way in which the natural scientists who lobbied for its creation had refused to include the social sciences in their proposal, had smuggled them in under the title of "other sciences," but in severely limited amounts.

More recently, the social sciences have received government support in amounts that are still small by comparison with the physical and biological sciences, but that would have seemed munificent a few decades ago. They still differ from the natural sciences in the political constraints they suffer, but much more as a matter of degree than of kind. All the sciences together are the political victims of their own success, in two ways: first, they come under pressure to contribute more directly to public purposes, and second, they are patronized or punished not for the knowledge they produce, but for the anticipated use to which it might be put.[18]

The pressure to contribute more directly to public purposes comes in part from the effort of politicians to collect on what they consider valid IOUs. Scientists speaking to scientists would urge the support of basic research by mission-oriented agencies (the Department of Defense by contrast with the National

Science Foundation, for example) in order to protect the freedom of scientists to deal with a variety of possible donors. But scientists speaking to congressional committees would naturally argue for the utility of basic research, at least within a relevant discipline, to the practical mission of the executive agency. In the eyes of a basic investigator, of course, any fundamental advance in science might be relevant to almost any mission, while relevance was defined more narrowly by congressmen accustomed to strict construction of statutes by auditors and lawyers. The amendment inserted by Senator Mike Mansfield in the Military Procurement Authorization Act of 1975 required the military services to support only such basic research as had "a direct and apparent relationship to a specific military function or operation." The effect turned less on verbal definition than on the general intent of the legislative history; although the amendment is no longer law, it stills puts a substantial damper on the discretion of military contract officers to contribute to the nation's basic research program.

Even more limiting to the basic research program was the growing tendency in Congress to shift funds in the direction of applied research. This was like the earlier move of private foundations to favor "problem oriented" rather than strictly disciplinary research.[19] Those members of Congress who favored the shift were not antiintellectuals or enemies of science; the Research Applied to National Needs program was pressed on the National Science Foundation by its warmest sponsors in the House Committee on Science and Astronautics, including Representative George P. Miller from California and Representative Emilio Q. Daddario from Connecticut. And such shifts were not forced on the scientific community entirely from the outside; one may recall Edward Teller's complaint that the United States had put proportionally too much emphasis on basic science by comparison with technology,[20] or the great numbers of biological and medical scientists who joined the lobby led by Mary Lasker in support of research dealing with specific diseases, or the ecologists and environmental scientists who linked their research not only to technological development but also to political action.

The danger of political constraints on science now comes not so much because politicians disapprove of the methods of science, but because they take seriously what some scientists tell them about the way in which scientific discovery leads to practical benefits. Once they believe that, it is inevitable that they should try to control practical outcomes by anticipating the effects of research, and manipulating it in one way or another.

In its simplest form, this manipulation consists in voting to support or to abolish research work leading toward intended purposes. Thus at one time in dairying states, research on the economic and nutritional aspects of the use of oleomargarine proved to be impossible in state universities,[21] and thus Congress may increase or decrease support for various fields of research in the light of its estimate of their relevance to specific purposes.

Less crudely and more indirectly, a legislature may seek to alter the structure of government organization in order to favor or punish research with policy implications. When Congress thought, in the fading years of the New Deal during the Second World War, that the National Resources Planning Board's research was leading toward economic planning, it abolished the board. When some congressmen thought that the development of ocean resources was being neglected, they tried to legislate the creation of a special oceanographic unit in the Office of Science and Technology of the Executive Office of the President.

Finally, and still more indirectly, legislative action may undertake to push scientific research directly into political issues, without knowing in advance just how the research will come out. In recent years Congress has by statute required the National Academy of Sciences to undertake more than two dozen major studies designed to answer policy questions. Some may have been voted by members of Congress who did not know what answers they wanted. Others had more obvious motives, as when Representative Jamie L. Whitten of Mississippi, unhappy about restrictions on the use of DDT in controlling the boll weevil, got a statute enacted to require the Environmental Protection Agency to contract with the Academy for a study of its own procedures.[22]

Similarly, many statutes now require that executive agencies contract with research firms to evaluate the quality of their programs; such evaluations may be based on the most modern techniques of social science, but they put a political strain on the ethics of many contractors, who are torn between the obligation of objectivity and the desire to qualify for another contract.

And finally, we move into the stage where science joins the demand for maximum feasible participation in democratic processes. Public interest groups engaged in criticizing or opposing official policies plead for congressional appropriations to support their activities, and those dealing with technological issues now look forward to the new "Science for Citizens" program of the National Science Foundation. This program, which the Senate seeks to expand from modest beginnings, would pay for the work of not-for-profit groups to employ scientists to work on public policy issues. The program is looked on with some apprehension by some members of Congress who fear it will, for example, fund the lobbying by public interest groups against nuclear power plants.[23]

In all these ways scientific research is involved in political controversy, and comes under political constraints. But it is important to note that none of them is the result of an effort by politicians to constrain science in the interest of religious dogma or political ideology. It may be good for the morale of a scientist who is caught up in political difficulties to make speeches recalling Galileo's suppression by the Church, or Lysenko's support by the Party. But in contemporary America, the main problem is just the opposite. Politicians have been persuaded by scientists that political issues can be solved by scientific methods, and hence use political power to increase the proportion of research work that is devoted to applied problems, to increase the amount of support that is given to institutions and programs for political reasons, and to drag (or welcome) scientists into political activity. All this may work against the best interest of basic research, which must be supported on a longer time perspective and with different motives, and for highest quality must depend on the work of a smaller proportion of interested talent. But it is not moving in the direction of thought control in the interest of a coherent ideology, but rather in the direction of dissipation of the energy of scientific institutions by too much involvement in the specialized practical interests that are the stuff of politics.

Bureaucracy and Executive Power

With the shadow of Watergate still hanging over our political memory, it is tempting to blame any political constraint, especially if it affects science or universities, on an overly powerful Executive controlling a disciplined mass of faceless bureaucrats. This conventional diagnosis is particularly inappropriate when dealing with issues in which scientific and political problems intersect, for four obvious reasons, in all of which the United States is nearly unique even among Western democracies.

1. In the United States' career civil service, scientists have higher status than general administrators. Two-thirds of the supergrade civil servants have come up from scientific, technical, or professional backgrounds. Unlike the corps of generalists or administrative lawyers that dominate the civil services of Great Britain or Western Europe, the United States' civil service has no unified administrative corps dominating its departments, and under the top political appointees authority is shared by career officers and short-term appointees from universities and private business.
2. Through specialized committees, Congress maintains oversight over the specialized bureaus rather more intimately than can the president's staff, and regularly amends executive proposals and especially executive budgets. In democratic parliamentary systems, there is no equivalent control in such specialized detail. Hence scientists in the United States, both within the government and outside it, have relatively free opportunity to know what goes on within official circles, and to influence congressional action on it.
3. The Executive does not have control over the organizational structure or the personnel system of the government, and the Congress manipulates both as a means of controlling policy.
4. Within the Congress itself, the individual members and committees do not permit the existence of any leadership (on the basis of party or otherwise) that can control the legislative proceedings, in the way that a Cabinet does in the classic parliamentary system. It is this kind of political discipline within the legislative body that makes possible in some democratic countries the existence of a strong chief executive, and the maintenance under him of a strong bureaucracy.

In its effect on the political status of science, the loose American system is the opposite of a strong bureaucracy. Executive authority lacks the authority not only to give the answer to a political issue, but to define the question. Science is seen as the most reliable basis for solving problems, but scientists are asked to answer not only specific questions to which their discipline may have the key, but broad questions of policy involving other specialized disciplines and a vast range of unscientific value judgments as well. In this setting, it does not matter that the typical complex issue of policy cannot be answered by a single scientific specialty; a congressional committee may insist that a nuclear physicist give it the word on arms control, or a biologist on the morality of abortion. This is an inviting opportunity for an eminent specialist endowed with high moral zeal and no sense of humor.

This system is very good at dealing with narrow and highly specialized problems, but less satisfactory with broader issues involving the coordination or reconciliation of competing values. That shortcoming is apparent not only when science undertakes to help solve issues of general policy, but also when it tries to deal with policy for the support of science. The peer review system, for example, works very well at deciding on project grants within the limits of a given set of universities and institutions, and a given balance of support as among scientific disciplines. It is incapable, however, of dealing with the broader issues that arise in proposals for institutional grants, of deciding between support of universities and special research institutes, or of determining how much money should go to chemistry as against biology or economics.[24] By default, such decisions are made (or recommended) by the staffs of congressional committees or of the Office of Management and Budget.

Science as a critical and skeptical mode of thought, more interested in pragmatic tests than in verbal theories, has probably been the main influence in dissolving traditional authority and in emphasizing the authority of the specialist, in all aspects of American life. When a university promotion is influenced by a numerical count of scholarly papers, when a government agency grants financial support in proportion to items that can be counted rather than on a general estimate of excellence, or indeed when civil service appointments are made by "objective" tests or the president must control his departments more by budget allocations based on statistical criteria than by loyalty to general policies—in all these ways we may observe the American overemphasis on the specialized and the quantifiable. And a good thing too, in my opinion, if we must go to one extreme or the other.

But in political affairs the extremes are just what it is important to avoid. To deal with the big problems of modern society requires the intellectual contribution of scholars and public officials who can take comprehensive and integrated (the currently fashionable word is "holistic") views. If they are to have a chance, it also requires institutional systems in universities, in business, and in government that encourage the broad and integrated view, and the long-term perspective on which it must be based, and the respect for the fundamental values of civilization on which its institutional structure must depend.

The protection of scientific inquiry from improper limitation by governmental authority does not require scientists to renounce an interest in public policy, or the rest of us to adopt a romantic notion that modern problems can be solved either by a return to primitive values, or by popular votes unaided by expert opinion. Facts and values (like policy and administration) are far too intricately mixed, far too reciprocally dependent on each other, for us to guarantee the autonomy of science (or the integrity and impartiality of the civil service) by forcing it to renounce a role in the great public issues of the day.

It helps, perhaps, to make a clear distinction in our thinking among science, scientists, and scientific institutions. Science is a mode of thought that it is important to distinguish from other approaches to human problems, even though it profoundly affects all of them. Scientists are citizens too, and ought to be encouraged to participate fully in politics, to which they may make a unique contribution as long as they make clear the limits of their competence as scientists to answer unscientific questions. But scientific institutions, including universities, may be under a special obligation, if they ask for governmental support, to emphasize the difference between their role and function and those of governmental agencies.

It is on a clear understanding of this difference that the universities' case for generous support and freedom from political constraint must depend. The universities can advance science and educate scientists, and science and scientists can then make great contributions to the general welfare, only if the universities are supported on a steady and long-term basis, and are free from political interference, including the benign type of interference that consists in requests to undertake work that is more suitable for a consulting firm or a government agency.

The protection of basic research against constraints is not mainly a matter of protecting one unitary force, the university, against a superior one, the government. If the institutional worlds of science and of politics have been badly mixed up over the past quarter-century, it has been less a case of violation by superior force than of mutual seduction. To work toward a future relationship of more adequate support, it will not be enough for university presidents to state correct policies. It will be even more important for the university scientists themselves to understand how they got into their present situation, as a guide to knowing how to get out of it.

The problem has not been one growing exclusively out of the demands of government; as the record of private philanthropy suggests, it has rather been one of a growing disposition on the part of many organized sources of support (and of the scientific community's more aggressive salesmen) to emphasize practical results rather than basic research and education. It has not been the assertion of political ideology or religious dogma over objective science; rather it has been the overreliance of political leaders on the fragmented authority of scientific specialties, and on the reductionist methods that they have popularized. And it has certainly not been the result of powerful executive or bureaucratic authority; the same overspecialized patterns of thought that afflict the academic community are built deep into the institutional structure and incentive systems of the Congress and the administration alike.

A Start toward a Strategy

In planning a strategy to reduce undue political or bureaucratic constraints on research, the first question may be whether to emphasize the similarities or differences between the world of scientific research and other parts of society. It is tempting, of course, to argue, and it is certainly partly true, that science, research, and indeed the world of academic and intellectual affairs generally are quite different from the mundane activities that may properly be expected to obey bureaucratic regulations. The practical trouble with this argument is that too many administrators of research and educational programs have been caught behaving like business managers generally, and, besides, the posture of intellectual superiority is not an ingratiating one to be assumed in a congressional committee hearing.

It is indeed possible that the political resentment against such a posture (even if the posture is more imagined by the beholder than really assumed by the scientist) has changed the nature of the relationship between governmental grantor and scientific grantee. It seemed to some during the 1950s that the distinction between governmental agencies and private research institutions was a line of defense that could continue to be held legally and administratively in order to prevent legislative and bureaucratic meddling in scientific affairs. At that time the public versus private distinction seemed a reasonably secure defense, unlike the symbolism of the corporate form of organization, which had been adopted when the government corporation was created in the 1920s to give more flexibility and discretion in the spending of government money.

But now the experience of the government corporation, which soon came to have as much red tape tied to it as ordinary government bureaus, is being repeated in the administration of government grants and contracts with private institutions; legislative conditions and administrative auditors can follow the taxpayer's dollar. Indeed, the atmosphere of the relationship has soured enough so that it may be an actual disadvantage, while facing congressional committees, to be a contract administrator representing a private institution rather than a regular civil servant. Whether or not this is so, it is clear that the same trend of imposing constraints on the spending of government money has operated in the field of contracts and grants to private institutions that formerly operated with respect to grants in aid to the states. It is obvious today that a university president who tried to assert, as a matter of legal right, that constraints should not be attached to the terms of research grants would look as silly as a state governor who still argued for states' rights in pre-Civil War terms.

But what cannot be defended as a matter of legal right may be as a matter of functional necessity. The extent to which authority is delegated in the expenditure of federal funds, or controlled in detail by congressional committees or administrative officials, varies tremendously from function to function. Scientists and educators may well argue, as I would, that it makes no sense for government to grant them money for their essential purposes, and then attach conditions to the grant that make it wasteful or ineffective. The mere fact that a program is supported by public funds does not necessarily require constraints in unwise detail, although it may often do so. If Congress is persuaded that the direction of a research program requires as much professional independence as, say, the direction of the Federal Reserve System or of the Joint Chiefs of Staff, or any other function that they consider professionally complex and important to the national interest, it will find ways to delegate in ample degree. Even some state legislatures manage to do so with respect to their state universities.

The functional approach would allow scientific institutions to avoid the appearance of pleading for unique privileges on the basis of elite and esoteric status. In practical terms, it would put them in alliance with other elements in society which might see some common interest in maintaining a degree of autonomy in a pluralist society. Some such allies would be in private institutions—business corporations or philanthropic foundations that are now willing to admit the need for government regulation, but would prefer to have it in broader and more flexible terms.

More important would be parts of government itself, for the same problems that arise in the relationship of government to private institutions arise also in the relationship of central control agencies like the Office of Management and Budget with operating agencies and bureaus. Both central control and operating agencies might be persuaded that detailed constraints are not only counter-productive but fail to deal with the broader and more important policy issues. If the General Accounting Office could make the change in that direction that it has accomplished over the past decade or two, any miracle can happen. And congressional committees, aware of the extent to which their specialization has obscured major policy issues rather than controlling them, and noting the effects of such an integrating reform as that of the Congressional Budget Office, and the potential of an effort like that of the Office of Technology Assessment, may yet see the need for focusing their attention on major issues and delegating more responsibility for details. It would be naive to expect instant results from such an approach, but even more naive to rely on the approach of promising results that basic research cannot deliver, and asking politicians to appropriate money for grants without then checking up on the results.

As long as congressional and executive staffs have incentives to react only to highly specialized and short-term pressures, no one in the political world can do much to promote the case for the sustained support of science for its fundamental and enduring values. The most effective (though perhaps the most difficult) first step toward a long-range and integrated policy of governmental support for science and for universities might well be to work for a more disciplined system of responsibility within the Congress as a whole, and a more competent corps of generalist administrators (by no means excluding scientists) who could give the planning system of the government more integrity and continuity.

But whether or not this approach would work, it would have the merit of discarding an approach based on an obsolete political mythology. It is bad enough for generals to plan to fight the last war. By now, scientists should understand that if they need to muster political support, it is not in defense of Galileo.

REFERENCES

[1]Vannevar Bush, *Science the Endless Frontier* (Washington, D.C.: U.S. Government Printing Office, July, 1945), report submitted to President Truman. The report should not be blamed for all political tactics since followed in its name.

[2]The case has been argued by officers of scientific societies and educational associations, and most recently in "An Overview: The Federal Government and the Major Research University, Indispensable Partnership for Excellence," in *Research Universities and the National Interest: A Report from Fifteen University Presidents.* To be published shortly; pre-publication draft made available by the Ford Foundation.

[3]These attitudes were responsible for the fate of the recommendations of the Science Advisory Board of 1934–35. The story is told in Lewis E. Auerbach, "Scientists in the New Deal: A Pre-War Episode in the Relations Between Science and Government in the United States," *Minerva* 3 (Winter 1965): 457–482.

[4]Cf. Daniel Bell, *The Coming of Post-Industrial Society* (New York: Basic Books, Inc., 1973)

[5]André Mayer and Jean Mayer, "Agriculture, the Island Empire," *Daedalus* (Summer 1974): 83–95.

[6]In addition to Bush himself (still head of the Carnegie Institution of Washington, which served as OSRD headquarters) the private foundation world contributed to the preparation of the Bush

report through committee members such as Warren Weaver of the Rockefeller Foundation, Henry Allen Moe of the Guggenheim Foundation, and others.

[7]Nathan Reingold, "The Case of the Disappearing Laboratory," *The American Quarterly*, 29 (Spring 1977): 79–101.

[8]Abraham Flexner, for example, as his autobiography makes plain. See *Abraham Flexner: An Autobiography* (New York: Simon and Schuster, 1960), pp. 274–275.

[9]As a stage in the evolution of the idea of a science foundation, one should not forget the short-lived National Science Fund, set up shortly before World War II at the initiative of Herbert Hoover to channel private funds into the support of science.

[10]The experiment of paying for consultants to cabinet members had been made in the early years of the New Deal by the Public Administration Clearing House with the aid of a grant from the Spelman Fund of New York.

[11]This point was argued in my *Government and Science* (New York: New York University Press, 1954), chap. 3.

[12]Bruce L. R. Smith and Joseph J. Karlesky, *The Universities in the Nation's Research Effort* (New York: Change Magazine Press, 1977), pp. 192–196.

[13]This story is told in a forthcoming study of the General Accounting Office by Frederick C. Mosher, of the University of Virginia.

[14]Smith and Karlesky, *op. cit.*, pp. 26–27, 187, 210.

[15]"An Overview," *op. cit.*

[16]Political questioning about DNA followed the Asilomar Conference; the Cambridge City Council inquiry depended on the initiative or advice of Harvard and Massachusetts Institute of Technology scientists.

[17]Tax-Exempt Foundations, Report of the Special Committee to Investigate Tax-Exempt Foundations and Comparable Organizations, House of Representatives, 83rd Cong., 2d Sess., H. Rep. 2681 (1954), pp. 17–19, 56, 60, 67, 73, 200.

[18]For a survey of the recent relations of the social sciences to government, see Charles Frankel (ed.), *Controversies and Decisions: the Social Sciences and Public Policy* (Russell Sage Foundation, New York, 1976). See also *Social and Behavioral Science Programs in the National Science Foundation*, final report of the Committee on the Social Sciences in the National Science Foundation, Assembly of Behavioral and Social Sciences, and the National Research Council Washington, D.C.: National Academy of Sciences,1976).

[19]E.g., *The Report of the Study for the Ford Foundation on Policy and Program* (Detroit: The Ford Foundation, 1949).

[20]Edward Teller, "The Evolution and Prospects for Applied Physical Science in the United States" in *Applied Science and Technological Progress*, Report to the Committee on Science and Astronautics, U.S. House of Representatives, by the National Academy of Sciences (Washington, D.C.: U.S. Government Printing Office, June, 1967), pp. 365–397.

[21]Charles M. Hardin, *Freedom in Agricultural Education* (Chicago: University of Chicago Press, 1955), pp. 119–121.

[22]"EPA Study: National Academy Set to Serve Two Masters," *Science*, 185 (August 23, 1974): 678, 680.

[23]"NSF Authorization Coming up," *Science*, 196 (June 24, 1977): 1423.

[24]This deficiency was recognized in 1965 by the report of the Wooldridge Committee, Biomedical Science and Its Administration: A Study of the National Institutes of Health (Washington, D.C.: The White House, February, 1965), pp. 10–13, 208–211. This was tacitly admitted by the series of studies by the National Academy of Sciences of the financial needs of various branches of science; no effort was made to put the several disciplines in priority order, or give them comparative weights.

WALTER P. METZGER

Academic Freedom and Scientific Freedom

It is striking how often references to "academic freedom" crop up in the current debate over the regulation of scientific inquiry, an ancient issue newly thrust into public consciousness by the alleged abuses of biomedical experiments and the apocalyptic hazards attributed to recombinant DNA research. Many of these references to academic freedom, it is fair to say, have little or no substantive significance. Some are merely the products of verbal habit; a debate waged in large part by academics is likely to adopt the idiom of academics, and no term in academic parlance is as readily summoned or as freely used. Some are offered for the sake of argument, to add the glitter of a favored symbol to a viewpoint that might otherwise lack appeal. But by no means should all of these citations be classed as merely ceremonial or polemical. Quite often, in the repeated invocation of the name and authority of academic freedom, one detects an expository motive, a desire by persons in moral doubt to be guided by a presumably related and potentially illuminating body of experience. On the technical issues in dispute—how human subjects may be protected from indignities or the environment from harmful laboratory emissions without hampering innocuous inquiry—few scientists may be inclined to look for guidance beyond their disciplines. But on the normative issues in dispute—whether the values of privacy and a safe ecology are more compelling than the value of pursuing truth, whether the long-term benefits of science should be sacrificed to the public's instant fears, whether scientists have obligations to society other than those of discovering, validating, and dispensing knowledge—many appear to be eager to receive instruction from a broader range of precedents and rules.

Ever since research became a major academic function, the principle of academic freedom has incorporated freedom of inquiry as one of its critical components. From this fact grows the expectation that the competing claims of freedom and responsibility, of science moved by its own imperatives and science bent to social purposes, have already been faced under this broad conceptual umbrella, and that the wisdom gained from such encounters has been embodied in readable files and texts. The frequent allusions to academic freedom thus appear to express the hope or confidence that this ethical domain, if it has not been fully charted, has at least been partly traversed.

This paper takes its cue from the rhetoric of serious intention. It asks *what does academic freedom, as conventionally defined and defended in this country, offer by way of precept or example to modern science in its moral plights?*[1] The first section undertakes to review the definition that has come to prevail in America, and that has been implemented through university regulations, professional campus investigations, arbitral decisions and collective bargaining contracts, and professorial-administrative pacts. The second section, which discusses the relationship of academic freedom to scientific freedom, evaluates the didactic benefits that can be drawn from this theory and praxis by those who wish to know how science can be free and at the same time rendered uninjurious.

I

A definition of academic freedom does not take hold and flourish adventitiously; it must be attuned to the key features of a particular academic system. That such a consonance is a requirement of success is aptly illustrated by the fate of a once esteemed definition, which was exogenous and ill-fitting in this country. Throughout the latter part of the nineteenth century, the very concept of academic freedom bore a German trademark. It was mainly from Germany that the world derived the notion that academic freedom not only is relevant to a university, but is absolutely essential to it, the one grace it cannot lose without losing everything. Only in the Germany of that period was academic freedom enshrined in a political constitution, eulogized in official speech and ceremony, and incorporated into the artless braggartry by which nations presume to exalt themselves. There academic freedom advertised itself through world-renowned scholarly accomplishments. Every striking achievement of German *Wissenschaft*—whether the flowering of speculative philosophy at the expense of dogmatic theology, the incursions of biblical scholarship into areas traditionally guarded by "No Trespass" signs, or the emergence of laboratory science as an alternative and rebuke to armchair theorizing—was given and taken as a testimonial to the power of academic minds left free. In this period, thousands of American graduate students flocked to the German Olympus for advanced instruction, and their sojourns, later recalled as trysts, lent an element of nostalgia to their penchant for academic emulation. Most particularly did Americans admire the German concept of academic freedom, which comprised a trinity of principles, *Lehrfreiheit*, *Lernfreiheit*, and *Freiheit der Wissenschaft*. For decades, these special terms were as common to an American essay on this subject as French words to a culinary dictionary. But this period of imitation did not last. After the turn of the century, American academics began to abstain from German passwords; by the advent of the First World War, they had fashioned a theory of academic freedom that was unmistakably their own. In part, their abandonment of one of the foremost gifts of *Deutschtum* can be attributed to their rising alarm over the Kaiser's military power and to an upsurge in native academic pride. But in the main, the German model was forsaken when American academics came to realize that the categories it embodied did not fit the academic system they had devised.[2]

The wonder is that they had ever thought otherwise. Perhaps one explanation of their earlier infatuation is that words ending with the suffix "freiheit," when translated into English could be easily misconstrued. Literally, *Lehrfreiheit* meant "teaching freedom," and when garbled by exact translation, could be taken to mean simply the absence of classroom censorship. In Germany, its meaning was more legalistic and precise; it referred to the statutory right of associate and full professors, who were salaried government officials working in universities supported by the state, to discharge their academic duties outside the bureaucratic structure of command that encompassed other civil servants. It meant that they could determine the content of their courses and impart the findings of their research without seeking ministerial approval or fearing ministerial reproof; it meant that, within a certain functional sphere, the restiveness of academic intellect would be protected from the obedience norms of hierarchy. But it did not protect professors outside that sphere. Although the privilege of *Lehrfreiheit* was deemed exceptional (it was not accorded to high councilors of state with whom professors were nominally ranked, nor to secondary school teachers from whom they were sharply set apart), it was also seen, in principle, as quite limited (it did not exempt professors, as civil servants, from adherence to a disciplinary code that required loyalty to the state, and it did not protect them, as private citizens, from the conformism demanded by the regime).

Academic freedom, as the Germans defined it, was not simply a professorial prerogative. Indeed, "Akademische Freiheit" generally alluded, not to the teaching freedom of professors, but to the learning freedom of students, or to *Lernfreiheit*. To an American, that term could be taken to signify a course of study filled with options and broad in range. In Germany, it meant much more than that; it amounted to a disclaimer by the academic institution of any authority over the student save that of qualifying him for state examinations and for culminating degrees. Having relegated character training to the family, general education to the *gymnasium*, and residence halls to the mists of history, the German university faced the student as a producer and purveyor of knowledge, not as a landlord, a custodian, or parent surrogate. And the German student, liberated from grades and classroom roll calls, required to find his own lodgings and diversions, free to move from institution to institution sampling academic wares, presented himself to the university as an independent being, not as a tenant, a neophyte, or a ward.

Finally, in imperial Germany, no definition of academic freedom was deemed complete unless it included a *tertium quid*—the right of the academic institution, under the direction of its senior faculty and their elected officers, to manage its immediate affairs.[3] A key component of *Freiheit der Wissenschaft*, academic self-government was presumed to link and guarantee the other freedoms. To protect the professoriate from ministerial dicta and ecclesiastical intrusions, the university, it was thought, had to be dissociated from public administration and allowed to live a corporate life apart. Faculty control of the university, in turn, was thought to depend on a large measure of student freedom, on the absence of guardian duties that would have required a large administrative apparatus and would have overtaxed the energies of the learned guild. In Germany, collective freedom and individual freedom were occasionally regarded as one, and always regarded as inseparable.[4]

In the closing decades of the nineteenth century, American academic reformers understood these concepts to be very well suited to their purposes. Those who sought to academize research were wont to repeat the catechism of *Lehrfreiheit*: universities exist to discover truth, truth is discovered by unending quest, questing must be free of ideological prescriptions. Those who opposed the traditional regimen of the college—its fixed curriculum, its enforced devotions, its stipulated if often unobserved asceticism—raised the grand banner of *Lernfreiheit*. And some American professors, citing the system they had observed as awed expatriates, agitated for greater faculty control.[5] As long as the borrowed phrases served merely as rallying cries, they were not closely tested for their congruence with the academic system that was taking shape. Once certain reform objectives were achieved, however, Americans did begin to perform that test and to relinquish irrelevant conceptions.

The first importation to be surrendered was the idea that broad-ranging student freedoms were an essential ingredient of academic freedom. If ever the German let-alone philosophy might have prevailed, that time had passed when the renovators of the system decided to keep the undergraduate college, either as a free-standing entity or as part of the university, and to retain its residential character. For a time it had been uncertain whether this part of the Anglican heritage would survive. Two centuries of experience with college dormitories had given them an unsavory reputation as breeding grounds for student vices and as staging areas for town assaults. But, as reformers discovered, there was a large and insistent demand for a collegiate home away from home—on the part of the student who delighted in the pastimes and comradeship it afforded, on the part of parents who sought a supervised abode for their callow offspring, on the part of alumni whose sentimental attachment to a part of the university (whether to a class or team or domus) had greater cash convertibility than did their loyalty to the complex whole. It was the triumph of the dormitory, as well as of the eatery, the infirmary, and the playground erected with it, that foreclosed the possibility of treating American students in the German mode. Even if *Lernfreiheit* could have made its way against the Calvinism embedded in our culture and the legal acknowledgement of in loco parentis norms, it could not have overcome the managerial solicitude stimulated by the heavy investment in what came to be known as "the collegiate way of life". Later on, the rise of a student "liberation" movement and a broadening of the constitutional guarantee of due process would combine to eliminate many parietal rules and curb the arbitrary enforcement of student discipline. But by then it would be much too late to resurrect *Lernfreiheit*. In America, academic freedom would ever be a one-class privilege; student freedom, even when it became a cause, would be generally regarded as something else, and by most academics as something less.[6]

American theorists of academic freedom never so thoroughly repudiated *Lehrfreiheit*. Yet the better they came to understand this concept, the more clearly they saw that it pertained to an academic and legal order that did not tally with their own. In this country, the field of higher education was not monopolized by the state, but shared by private as well as public agencies; professors were not officials of the state, but employees of the governing boards of academic institutions; universities were not oases of freedom scattered over an autocratic landscape, but the almost indistinguishable parts of a generally democratic society. Temporarily useful as a rousing catchword, *Lehrfreiheit* belonged to a world too strange for it to be accepted intact as a working principle. Gradually it was reinterpreted and revised (in ways that we shall presently examine). By contrast, the third German component of academic freedom, *Freiheit der Wissen-*

schaft, experienced a modest upsurge in popularity after the turn of the century, only to fall afterwards into a precipitous decline.

In America, again for reasons particular to its system, academic freedom was destined not to become a collective noun. It would stand for personal freedom, the freedom of the academic; it would not stand for collective freedom, the freedom of the academy, except in limited circumstances where a threat to the autonomy of the institution (say, a legislated loyalty oath required as a condition of employment) bore immediately and directly on the academic. The reason for this shift in emphasis could hardly have been that the American university was singularly untroubled by external threats. If it had nothing to fear from a bureaucratic monarchy or an established church, (at least not since the American Revolution), it had much to fear from a host of assertive publics that treated higher as well as lower education as fields in which they might legitimately impose their will. The problem of the all-too-pregnable university in an all-too-predatory society was real and worthy of profound anxieties. But it was overshadowed in the minds of American academics by another problem, their insecurity in the face of the discretionary power of resident governing boards and their appointed deputies. This insistent, inescapable problem was *sui generis*; no other major academic system brought the lay world so deeply into the academy or set up so elaborate a machinery of on-site administrative control. In their preoccupation with this problem, American academics came firmly to believe that the primary concern of academic freedom was with what happened *in* a university, not with what happened *to* a university—that an offense against academic freedom was essentially an inside job.

An American theory of academic freedom did not become crystallized in this country until 1915, when Arthur O. Lovejoy of Johns Hopkins and E. R. A. Seligman of Columbia, assisted by a committee of academic luminaries including Frank A. Fetter of Princeton, Roscoe Pound of Harvard, and Richard T. Ely of the University of Wisconsin, wrote "The General Report of the Committee on Academic Freedom and Academic Tenure" on behalf of the newly established American Association of University Professors (AAUP). For the student of academic freedom, this short philosophical essay, to which a set of "practical proposals" was appended, is extraordinarily revealing of the mind of a profession that was passing from a traditional to a modern footing, and that was seeking a credible ideology to protect and facilitate that change. It points up the difficulties faced and the critical choices made in adapting a legacy of imported concepts to native problems and conditions. Above all, it provides a rationale for academic freedom, the first fully developed one indubitably of the New World, a rationale that future generations were to live by, albeit with some modification in emphasis and many refinements of detail. To consult this sixty-year-old treatise is thus to find the key to current meanings, as well as to discern the turning points in an important chapter of the history of ideas.

The pervasive influence of the problem of lay control on the minds of academics of this generation is clearly evidenced in this report. Indeed, concern over the presence of lay potentates within the walls shaped almost every significant line of argument. In the philosophical section of the report, Lovejoy and company posed a fundamental question: Do faculty members have the same obligations to their governing boards that employees ordinarily have to their employers? Around this question revolved the following ancillary questions. To whom does a university belong? In whose behalf do trustees hold their trust, and regents discharge their responsibilities? Who are the clients of professors and how do their respective claims affect the accountability of professors to their institutions?

Their case for academic freedom rested on the answers given to these fiduciary questions. First and foremost, the AAUP authors argued that members of a faculty were not the employees, but the appointees, of their boards and as such were entitled to as much intellectual independence from their appointing authorities as judges in the federal courts had from theirs. To support this reading of the faculty's employment status (which most lawyers at that time would have called erroneous and many administrators would have called vainglorious and absurd), they redefined the role of the employer. They argued that the trustees of private universities were not the agents of the benefactors, but the agents of the general public; that the regents of state universities were not the instruments of the existing public, but the instruments of future publics; that the academic profession serves not only an intramural client—the student who obliges the institution to offer sound instruction—but also two extramural clients—the public agency in need of expert guidance, and the general society in need of greater knowledge; that the latter clients have a stake in disinterested professorial opinion, stated without fear or favor, which the institution is morally required to respect; that the institution does a disservice to all the profession's clients, including students, when it punishes or threatens to punish an academic person for his views.

Such stipulations as to who owes what to whom were to become the very centerpieces of the theory of academic freedom in America. Not every theoretical work would draw on precisely the same authorities (these authors combined a market theory of truth taken from J. S. Mill with a concept of stewardship reminiscent of St. Paul); and not every theoretical work would employ precisely the same images and analogies. But all ratiocination on this subject, given the professorial anxieties generated by the asymmetries of the trustee system, would engage, to some extent at least, in this kind of inward-turning moral sociology.[7]

These professorial anxieties did not lead to the displacement of lay authorities and the installation of faculty members in their stead. Although in AAUP circles at this time a faculty syndicalist movement, led by the Columbia psychologist J. McKeen Cattell and drawing inspiration from both German and English examples, did gain considerable notoriety and a measure of vociferous support,[8] the radical reconstructionists were never strong enough to prevail, and the general report of 1915 was not written by professors with the most extreme guild complexes. Rather, it fell to the dominant Whig contingent of the AAUP, which was strongly represented on the academic freedom committee, to separate the idea of faculty control, which was alien and threatening to those in power and by no means enticing to all professors, from the idea of administrative constraint. Based on the principles of academic tenure and due process, such constraint could be reconciled with the language of college charters, would not saddle scholars with unwelcome duties, and was likely to meet with less adamant resistance. The idea of administrative constraint, which called for faculty participation in the prosecutory and judicial processes of the institution, was embodied in the practical proposals of the 1915 statement and in all subsequent academic freedom codes. The idea of faculty control, which called for faculty involvement in the legislative and administrative processes of the institution, was omitted from this document and in general from the creed of academic freedom. (Later, as insurgency receded, this idea was softened into a bid for joint arrangements—"shared authority," and more recently, "collective bargaining"—that sought to limit lay authority without uprooting it.)

Moderate in its approach to governance, the 1915 report appeared to expound the common sense of its subject, and thus appealed to the profession's and the nation's sober citizenry. Yet if this approach made academic freedom more respectable, it also made it more introverted and parochial. A sustained interest in faculty control might have brought the troubled relationship of the university to society into sharper focus. Instead, the attention of the American professoriate was turned ever more intensely to domestic issues. Trying to perfect an interior rule of law, the theorists of academic freedom concentrated on probationary limits and predismissal hearings, on written terms of appointment and notice of termination in due form. They did not concentrate on the political practice of turning regential posts into party spoils, on the legislative habit of voting academic budgets line by line, on the popular inclination to intrude local ethnic and national patriotic causes into the curriculum fixed by law. Such external issues were presumed to belong to the field of academic diplomacy, not to the field of academic freedom, and were thus consigned to presidents, not professors.[9]

Important as it was, the presence of lay authority in academia was not a template stamping inescapable solutions on every analytic problem. In fact, to one analytic problem of great importance, the members of the AAUP committee proposed very divergent solutions, at least at the start of their deliberations. Once student freedom had been jettisoned and collective freedom pushed aside, there remained a residuum to be grappled with: to what types or classes of freedom are professors entitled? Seligman, the chairman of the committee, preferred a classification based on forms of professorial expression, "freedom of inquiry, freedom of utterance (whether by written or spoken word), freedom of action." Lovejoy, who was chosen by the group to act as editor and who acted as editor-very-much-in-chief, preferred a set of categories based on types of professorial activity, "freedom of inquiry and research, freedom of teaching within the college or university, freedom of extramural utterance and action."[10] The key word in Lovejoy's formulation was "extramural." In 1915, the idea that academic freedom was present in activities not related to academic teaching and research was not yet conventionally accepted. A number of AAUP committee members, still arguing in the German manner, were of the opinion that academic freedom could apply only to professional pursuits. A few were even of the view that it could apply only to professional pursuits carried on inside the borders of the institutions.[11] Conscious of the practical difficulty of distinguishing between on-campus and off-campus research, sharing the progressive's inclination to release expertise from the ivory tower, Seligman was willing to accept the concept of extramural freedom in a narrow sense; he would extend the protections of academic freedom to words uttered off the grounds but in the course of duty. He found it difficult, however, to accept the notion that the public pronouncements of professors on matters unrelated to their competence should enjoy the same immunity or the same respect.[12]

After considerable discussion, Lovejoy prevailed upon his colleagues to relinquish every zonal ordinance. The final draft of the report brushed aside the suggestion that the campus fence should define the outer limits of academic freedom. And it labeled "undesirable" any effort to draw distinctions that would place the speaker of unscholarly words in jeopardy, while protecting the speaker who spoke with the authority of the chair. As finally defined, academic freedom was deemed to give a protective cover, not only to teaching and research conducted off the campus, but to the exercise of all the political rights vouchsafed to every citizen, including the right to take a stand on public issues and to support a candidate for public office. (Only on the right of a professor to run for public office did the report remain equivocal; this issue, left for "further study," gradually lost its tie to moral principle and became bound up in the narrow question of the duration of faculty leaves of absence.) As a result of these categorical decisions, civil liberty was assimilated into academic freedom. From that time forth, it was to constitute one of the three main parts into which this Gaul would be conceptually divided.[13]

Two facts of academic life weighed heavily in favor of the Lovejoy argument. One had to do with the urgencies of recruitment; in a society that valued free speech and broad political participation, a profession that was willing to inhibit either would be at a great disadvantage in attracting talent. The report touches on this concern when it declares that, since persons would not and should not be drawn into the profession "by the magnitude of the economic rewards it offers," it is all the more necessary that persons of high character and ability be enticed by "assurances of an honorable and secure position, and to freedom (to act) honestly and according to their consciences . . ." The notion that academic freedom was a kind of fringe benefit, something to be added to the contract in lieu of a generous pay scale, may not be the noblest argument that can be made for it. But the belief that a second-class civil status would fail to attract first-class intellects probably was and is correct.

The second influential consideration is suggested by the report's prefatory comment that all the academic freedom cases investigated by the AAUP in the previous year had concerned "the right of university teachers to express their opinions in their capacities as citizens."[14] Numerous calls for help from academics had been received by the association in its natal period, and Lovejoy, its chief organizer and guiding spirit, had stolen time from his philosophic labors to look into some of these tangled, bristling affairs. He was convinced by his campus investigations that, in diverse regions of the country, forthright professors became an endangered species when they ventured beyond the walls. In universities as disparate as Utah and Wesleyan, Montana and Pennsylvania, and on issues ranging from child labor and sabbatarianism to the deferences due an academic president, it was the off-campus, professionally unrelated utterance, and not the scholarly or pedagogical utterance, that got professors into trouble with their superiors and in some cases cost them their jobs.[15] The decision to stretch the principle of *Lehrfreiheit* to cover extramural conduct thus had a finding of fact to recommend it; it served to protect American academics where they proved to be most exposed.

Some recent commentators have suggested that there was yet a third reason for this extension—the weak protection then afforded by the law against threats to freedom of speech posed by the punitive employer. In that day and until quite recently, it was held by the courts that the government as employer (and, a fortiori, the private entity as employer) could extend or withdraw the privilege of employment on such grounds as it saw fit to use without raising a constitutional issue, let alone trespassing on constitutional rights.[16] At the same time, judicial interpretations of the academic employment contract had generally granted the academic employer broad discretion in determining whether (and, to a large degree, how) that bond would be dissolved.[17] Thus, as far as the law was concerned, the faculty member whose speech was displeasing to his employer could not be punished criminally but could be punished economically. Yet, as far as the faculty member was concerned, the threat of disemployment, if less horrendous than the threat of incarceration, was somewhat more probable and a whole lot more real. Consequently (so the argument goes), free speech was tacked on to the table of contents of academic freedom, there to gain a professional protection that was better than nothing but still second best. In company with this explanation, goes the judgment that academic freedom fared the worse for allying itself with civil liberties. It acquired a reputation for indiscriminateness, since it gave the ignoramus, equally with the knower, a valid claim to its protection. At the same time, it picked up an elitist connotation, since it sought to make academics safer than nonacademics when both were saying the same things. To the difficulty of convincing outsiders that professors deserved any special privilege was thus added the further hardship of explaining why professors, when they spoke as citizens, should be treated more delicately by their employers than (say) carpenters and janitors, in the same capacity, were treated by *their* employers, who might, indeed, be the *same* employers. Taking heart from the recent overthrow of the privilege of employment doctrine, some now argue that the academic profession should give over this tutelary burden to the judicial guardians of the Constitution, and thus not only relieve academic freedom of its liabilities but strengthen civil liberties as well.[18]

No one aware of the fortunes and misfortunes of academic freedom in this country would wholly disagree with this analysis. Having to defend civil liberty under the rubric of academic freedom has proved awkward and burdensome; the lack of a serviceable First and Fourteenth Amendment did greatly weaken that defense. But no one who examines the original and continuing rationale of academic freedom in this country would believe that extramural freedom was added to it as an extraneous item, faute de mieux. The addition did serve to fill a legal void, but not as a patch of strange material might cover a hole in an otherwise intact garment. Rather, it was the logical (if not inevitable) extension of the fiduciary arguments for academic freedom, the further-spun threads of a single cloth.

The assertion that the legal guardians of a university are the representatives of the whole society always carried with it an implicit corollary, that they could not put the stamp of the university's name on a disputed truth claim without being faithless to their social trust. This call for trustee abstention, which would later be broadened into a demand for "institutional neutrality," tended to erode the distinction between inner and outer speech.[19] Fundamentally, it did not matter whether a governing board, speaking for a university, took issue with an incompetent or soundly backed opinion, with a thought expressed in the public forum or in the campus lecture hall; what mattered was that a board engaged in the unseemly act of partisanship. In this sense, extramural freedom was of a piece with teaching and research freedoms; professors had to be protected in every role and in every forum so that the university might be protected from leaders who would use it to sponsor doctrine. And in this sense, too, not even the most sanguine assumptions about recent constitutional trends would warrant the removal of extramural freedom from the academic freedom list.[20]

II

We return to the expectation that prompted this excursion—the hope that a discussion of academic freedom might elucidate the limits of scientific inquiry. What in the canon has the capacity to be so illuminating? To note that it draws on the prestige of science, that it endorses freedom of inquiry, that it counts academic scientists among its intended beneficiaries is not to demonstrate that it casts any direct light. More would be said about its candlepower if it could shown that it clarifies the ethics of science, or that it has produced an informative record of cases touching on free inquiry, or that it accurately identifies the controlling forces that contemporary scientists have most to fear. But on these matters, as we shall see, the opposite is more nearly true, and for fundamental reasons. At the end, we come to a discouraging conclusion: the groping scientific conscience may take diffuse enlightenment from this fund of experience, but if it looks for a beacon, it will look in vain.

Let us consider first the ethical injunctions in this canon. It is important to note that the AAUP report posed not one but two kinds of ethical questions: not only "To what types of freedom is a faculty member entitled?" but also "To what limits on these freedoms is a faculty member obliged to submit?" The second question is surprising only to those who expect professionals in their thoughtful moments to produce completely self-serving ideologies. These professionals, in any case, did not; while they were hardly self-effacing, they did hold that an acceptance of limits, which they called "responsibilities," had to accompany, clarify, and justify an assertion of professional rights. As far as teaching was concerned, they conceded that a professor's right to speak his mind was limited by his professional concern for his students' minds. For their sake, he was under an obligation to discuss controversial subjects "in a fair and judicial manner," to temper instruction with discretion when accosting youthful preconceptions, to teach students to "think for themselves." As far as extramural utterances were concerned, these authors held again that the professor's right was limited by his professional duty to protect the dignity of his calling. To this end, he was under an obligation "to avoid hasty or unverified or exaggerated statements, and to refrain from intemperate or sensational modes of expression." Conscious that these maxims could be abused, they asserted that they should primarily be enforced by the "public opinion" of the profession or by the individual professor; only when their alleged violation took the extreme form of "habitual neglect of duty" or a "grave moral delinquency," and never when the validity of an expressed opinion was in dispute, did they concede that the governing board was competent to intervene.[21]

They did not, however, apply this limitary principle across the board. When it came to freedom of inquiry, these same academic ethicists, who refused to make the other freedoms absolute, saw fit to impose no limits whatsoever, even in the form of self-applied admonitions. In this area, they would admit no conditional "ifs," no cautionary "buts." They stated that "in all [the] domains of knowledge, the first condition of progress is complete and unlimited freedom of inquiry. Such freedom is the breath in the nostrils of all scientific activity." In the midst of so many restrictive clauses, this unqualified assertion, whatever its intrinsic merit, plainly stands out as an anomaly.[22]

It was an anomaly, but not an idiosyncrasy. Doubtless any group of scholars writing on this subject in this period would have said as much, which is to say, as little. Indeed, it is likely that most members of governing boards in most secular and in a good many religious institutions of higher learning would have agreed. An idea so close to being part of a consensus omnium would appear to reflect the spirit of the age, and this one, so straightforward, so free of doubt, would appear to reflect the spirit of a singularly placid and untroubled age. One observes that the kind of science visualized in the report is primarily an intellectual activity, working to uncover the general laws of natural and social phenomena. To the extent that it was also visualized as a manipulative activity, using controlled observation and experiment, it was seen as following a heuristic method, not as playing on objects and subjects in possibly unsafe ways. Least of all was it visualized as an instrumental activity, possessed of its own technology, which might give off toxic byproducts or in other ways generate social costs. For these authors (who were mostly philosophers and social scientists and perhaps the more inclined to define research as a purely ideational pursuit), there was no such thing as dangerous knowledge; whether there was such a thing as a dangerous way of gaining knowledge, they did not wonder or at least did not ask. One notes, further, that the report written in 1915 exudes a confidence in the redemptive power of scientific knowledge, in its unmitigated beneficialness for humankind, that is plainly pre-Hiroshima, even pre-Verdun. Though the great European war had started, shattering the optimism of social philosophers, these Americans had probably not yet grasped the disillusive lessons of the battlefield—that the chemistry that advances human life can also be utilized to destroy it, that the biology that cures disease can also be put to homicidal use. Scientific freedom, "complete" and "unlimited," echoes a belief in the idea of progress that had yet to be overthrown.

But it would be a mistake to attribute the unconditionality of this precept solely to the cheerfulness of its authors. The chances are high that they would not have put academic science on a tighter leash even if they had been brutally acquainted with the paradoxes of modern science. At least, so their later actions argue; when America finally went to war, these professors jumped at the opportunity to put their disciplines at the disposal of the generals and propagandists—on both sides of the water, knowledge of Mars did not deter Minerva from embracing him, when she was persuaded that she was doing so in a just cause.[23]

The best indication that the demand for unlimited inquiry was not entirely due to the innocence of its proponents is that it kept reappearing in the protocols of academic freedom long after the passing of the Edenic age. In 1940, the representatives of the AAUP joined the representatives of the Association of American Colleges, an organization of college presidents, to create (inter alia) a mutually acceptable map of the limits of academic freedom. After prolonged negotiations, the heirs to Lovejoy and Seligman agreed with their presidential counterparts to limit freedom of teaching by one injunction (against introducing irrelevant controversy into the classroom) and extramural freedom by other injunctions (a string of "shoulds" designed to raise the level of polemical civility among professors and to dissociate the institution from whatever they might say). Furthermore, the two sides agreed, after much backing and filling, that the standards laid down for extramural utterances could be used as criteria for disciplinary actions initiated by institutional administrations. But on the matter of freedom of inquiry, neither side had anything very new to say. The AAUP team, in the spirit of its forebears, got the presidents to agree that there should be "full freedom of research and in the publication of the results . . . ." The presidential team got the professors to agree to two qualifications to this sweeping license, neither of which would affect the content or the conduct of inquiry. One was that the time devoted by a faculty member to research should not impair "the adequate performance of . . . other academic duties"; the other was that "research for pecuniary return [should] be based upon an understanding with the authorities of the instituion." After twenty-five years, the codifiers of academic freedom could advance their thinking on the ethics of inquiry only to the point of worrying about the venial sin of moonlighting and the possible misuse of company time.

This steadfast insistence on freedom of inquiry, with no restraints or only trivial ones, reflects something more than an inveterate philoscientism. It should be borne in mind that the formulators of academic freedom doctrine were not trying to write a code of scientific ethics, nor even a comprehensive code of professional ethics; they were trying to persuade academic authorities not to play a directive role in certain spheres of academic activity, that is, they were trying to construct a partial code of managerial ethics. Their acceptance of "limits," however genuine, was always highly tactical, and bound therefore to be selective. In those spheres where the interests of the institution were such as to invite the oversight of its legal guardians, they saw the need for entering a quid that would induce the desired quo. Thus, in the spheres of teaching and public utterance, where students could be seen as the charges of the institution and the good name of the institution might be at risk, they were willing to buy a greater degree of administrative benign neglect with a pledge of professorial conscientiousness. But they took a different view of scientific inquiry, which served no immediate clientele, did not affect the institution's reputation except vicariously, and did not submit to purely local standards of evaluation. In this sphere, to have agreed to a pact of mutual limitation would have been to enlarge the managerial jurisdiction—to have implied either that, without such a bargain, administrations could rightfully intervene, or that, given such a bargain, administration has the duty to make sure that it was carried out. Here it seemed more appropriate not to engage in the process of give-and-take, safer to be silent about constraints.

That silence obviously makes these codes unhelpful to persons seeking to set the boundaries of scientific freedom. Do the practical applications of these codes to concrete cases offer an assistance not provided by their formal precepts? In the professional defense of academic freedom, the case approach serves a two-fold function: it permits specific controversies to be judged in the light of codified principles and it permits codified principles to be amplified and reinterpreted by the clarifying factuality of specific controversies. The second of these functions supports the hope that more can be found in the case files of the sentinels than appears in the products of the conceptualizers.

Alas, this hope does not bear fruit, as is evident from an examination of the cases handled by the leading monitor of academic freedom in this country, the large constabulary apparatus known as Committee A of the AAUP.[24] A review of the historical records of this body, which has unremittingly patrolled this area for more than sixty years, reveals relatively few alleged infringements of scientific freedom and even fewer authenticated complaints. Of confirmed infractions, there are only occasional and scattered instances: here a psychologist dismissed for venturing to survey the sex habits of his students; there a research team in an agricultural college drawing fire from the dairy industry in its state for declaring oleomargarine nutritious; elsewhere a professor in a Catholic woman's college expelled by the president reverend mother for writing a novel she regarded as risqué. (The word "scientific" is used here in its broadest connotation; were it used more narrowly to refer to the study of the phenomena of the physical world, the gleanings would be thinner still.) Out of a sample of 1,500 cases drawn from this 4,000 case collection, this author could find only a dozen or so cases of substantiated violation of free inquiry, and of these only two or three that bear even remotely on the issue of scientific freedom and its due constraints. The sifted result hardly provides enough material from which instruction for today's dilemmas can be derived.[24]

Most AAUP cases raise questions of procedure, not of substance; they are tenure cases pure, if not always simple. Of those complaints, validated or not, that question the "why" as well as the "how" of an academic nonreappointment or dismissal, the vast majority attribute the punishment of the complainant to personal animosities, gender or racial prejudices, disapproved participation in campus protests, and unpopular political associations or unpopular political beliefs. Those complaints that do allege attacks upon freedom of research tend to belong to the sheaf of unconfirmed suspicions. The bulk of these are assertions by junior faculty members that they lost out in their bids for tenure for reasons more closely linked to the ideological slant than to the scholarly or scientific merits of their published work. Whatever may be said about these assertions, they cannot be taken as hard examples of constrained inquiry. Such are the evidentiary requirements of the AAUP (and such are the closed-mouthed ways of senior faculty and administrators who pass judgment on tenure matters) that these charges are seldom susceptive to the minimal proofs required to propel a full-scale Committee A investigation. Languishing unconsummated in the files, these one-sided complaints (which may, for all that is known, be fantasies) have to be regarded as enigmatic gifts donated to the present by the past.

Why is the AAUP docket so sparse in freedom-of-research cases? Part of the answer lies in external causes. Only a small percentage of the American professoriate—one estimate puts it at ten percent—actually engages in research. Obviously, with so small a fraction of the population in the pool of potential complainants, there can be no higher incidence of complaint. It may also be true that scientists communicating with scientists are not very often overheard, or, if overheard, are not very often understood, by antagonistic laymen. Esoteric journals and technical argots, whatever other ends they serve, do reduce the amount of gnostic eavesdropping that may be used to punish scientific acts. Finally, those who publish works that disturb the world—they are not innumerable—tend to be associated with universities that praise them for their accomplishments and protect them from counterstrokes. Complaints do not often go to the AAUP from professorial enfants terribles who earn rapid promotions and are awarded honorific chairs.

But the demography and ecology of academic research, and its often protective inscrutability, do not wholly account for the paucity of cases bearing on freedom of inquiry. Another factor, one more internal to the operation, must as well be cited; the AAUP works with an interpretation of academic freedom that alerts its police forces to certain trespasses and screens out the rest. According to this standard interpretation, a violation of academic freedom is a drama in which an outspoken professor is the victim, a repressive governing board or administration is the culprit, the power to dismiss is the crucial weapon, the loss of employment is the telling wound. For all that it harmonizes with the givens of our academic system, this interpretation tends to discourage the submission of less theatrical scripts. It tends to pass over the dangers to scientific freedom that may lurk in quotidian operations, for example, the constraints that flow from the partisan allocation of institutional resources or from a pattern of departmental appointment that shuts out significant schools of thought. It tends, further, to neglect transgressions that are committed by nonofficial persons, for example, by students who may, by open or subtle pressures, deter a scientist or scholar from pursuing controversial research. From a mountain of heterogeneous experience, the AAUP thus tends to extract a selective caseload, and a certain volume of iniquity, which cannot be measured, is left unmined.

This is not to disparage the work of the AAUP. Championing its principles case by case, that body has given vindication, if not always adequate redress, to a host of academic men and women wronged in some fashion by their institutions. By naming and shaming unjust authorities, it has chastened a good many academic presidents who equate loyalty to Alma Mater with obsequious reference to themselves, and a good many governing boards that believe that the best way to deal with faculty contentiousness is to fire it. It has greatly reduced the possibility, not remote to well-stocked memories, of trustees judging the qualifications of professors in the light of their class obsessions and of administrators using the friendliness of potential donors as a test of what professors should be allowed to say. Its work remains invaluable; it adds measurably to the quantum of decency in American academic life. But its work, for all its merits, does not reach many of the perplexing issues associated with freedom of research.

In recent years, as a consequence of the changes in the conditions of academic inquiry, the doctrine of academic freedom and the protective efforts that flow from it have grown even less suited to this task. As long as most scientific research was not only housed but supported by universities, largely out of operating budgets, it could be regarded as adequately protected by the creation of a zone of immunity free from the dictates of the managers. But when scientific research climbed from the Edison to the Brookhaven scale of costs, and when the federal government began to pay the major part of the bill, a blunt "hands off," addressed to powers resident on the campus, became a less relevant and less practicable prescription. Increasingly for researchers in the behavioral sciences, overwhelmingly for researchers in the natural sciences, the chief source of support, and therefore of potential mischief, is no longer the academic employer, whose area of control was always bounded and whose urge to control was in time contained, but the agencies of the central state, which give and guide on a national scale and which constantly add to the facets of social life they would improve through conditioned disbursements. Already facing a system of scientific regulation in which universities set up research review committees and file environmental impact statements to comply with the guidelines of governmental funding agencies, the proponents of academic freedom hardly touch the complexities of the world around them when they speak of "complete and unlimited freedom of inquiry" and other such absolutistic formulae. Nor, in an age when the targets of research can be set by the thrust of the yearly congressional appropriations or by agency requests for research proposals or by research and development projects directly organized and supervised by the military services, do they communicate with the primary source of power when they hold their colloquies with trustees. Nor, finally, under this new and complex dispensation, do they fully grapple with reality when they define an attack upon scientific freedom as an attack on the livelihood of the scientist. This *can* be so: an academic scientist may be denied a federal grant, or refused a needed clearance, because of his political associations; a government scientist may be removed from his post for communicating bureaucratic secrets to the Congress or the press. But an attack upon scientific freedom may also take the form of a lucrative inducement by an arm of government to do research under auspices and for purposes that are concealed. Temptations may not be the equivalents of coercions in either a legal or moral sense; still, rewards may be more effective as a mode of control than punishments, since, by capturing volition, they undermine the capacity to resist.

It should be clear, from all that has been said, that academic freedom and scientific freedom are different species of freedom; the time has now come to state schematically what it is that makes them so unalike. The key differentia lies in this: academic freedom is the ideology of a profession-across-the-disciplines, the profession created out of the common circumstance of an academic appointment in a college or university and of the common duties and anxieties that this entails; scientific freedom is the ideology of the divers professions-in-the-discipline, the professions based on the regularized advance of knowledge in distinctive fields.[25] Each affects a different, though overlapping, constituency. The protections of academic freedom extend to all in the class called faculty members, whether or not they do research (most do not do research, but almost all are associated with a discipline through specialized study in their training institutions and through departmental affiliations in their career institutions). The protections of scientific freedom extend to all in the class called researchers, whether or not they are members of a faculty (depending on the field, a small or very large percentage will not be faculty members, but members of other private or public institutions or of that self-employed contingent called "free lance"). Each develops a different, if not entirely disparate, criminology. Academic freedom is wise to the ways of the harsh employer but lacks a theory and vocabulary for dealing with the external offender and the nonoccupational offence; scientific freedom, though it has not yet articulated a penal code, can be said to enjoin a wider range of misbehaviors—those committed by the funding state no less than those committed by the campus paymaster, those committed by employers seeking profits no less than by employers with eleemosynary goals. Like all typologies, this one runs the danger of being too disjunctive; real people do not dwell in the cells of abstract schemes. The supporters of these ideologies may not be so far apart as the ideologies themselves. Historically, they have often worked in tandem: the battles between evolutionary science and religious orthodoxy waged within and without the university are proof enough of this. Temperamentally, they have much in common, sharing the same edgy mistrust of lay authority, the same sense of being beset by nescient forces, the same belief that truth must be found through the clash of minds, rather than through the imposition of official edicts. And many individuals, by belonging to both the broad academic profession and the specialized scientific profession, maintain commitments to both ideologies that can be fastened together or even interfused.[25] But these affinities, which exist in the minds and social roles of the upholders, do not dispel, and should not be allowed to mask, the differences resident in the ideas. It is these differences that point to what may be considered the overall message of this paper: *scientific freedom needs its own theoretical formulation of rights and limits, and its own machinery and procedures for detecting and reproving an offense.*

Suppose the devotees of scientific freedom, under the auspices of the American Association for the Advancement of Science or a yet more inclusive body (which might be called the American Association of Professional Inquirers or the American Research Liberties Union) were to attempt to compose a philosophic statement comparable in sweep and depth and potential influence to that written by the advocates of academic freedom on behalf of organized professors many years ago.[26] In general, what matters would it touch on, and what matters, of absorbing interest to that prototype, would it put aside?

It would concentrate on the relationship of science to society, not the relationship of faculties to administrations. Extricated from the problem of campus governance, the issues of freedom of inquiry would be revealed in a variety of transactions that find no convenient niches in the theory of academic freedom—for example, those between scientists and industrial employers, between scientists and governmental employers, between scientific enterprises and local interests, between the scientific community and the patron state.

It would offer a reasoned defense of free inquiry, but it would not be constrained to do so by dilating on the nature of the university, the functional importance of professors and the forbearance required of trustees. Paying less attention to academic pride and precedence, it might explore more fully the ethos and methodologies of science, especially in their modern forms.

It would not flinch from acknowledging that science has obligations to society over and beyond that of advancing knowledge. Where the treatise on academic freedom uses the name of society simply as a counterweight to lay authority, a comparable treatise on scientific freedom would give concrete meaning to that abstraction by calling it "the public interest" and by defining it as a set of demands for candor and circumspection with which the scientist is obligated to comply. Thus, it might insist that a scientist working for a private company has the duty to inform the public of dangers inherent in a newly developed product, though such a "blowing of the whistle" might be construed as an act of disloyalty to the employer. In the same vein, it might insist that every scientist must be solicitous of the health of all who are touched by his research—not only subjects and coworkers, but neighbors and consumers. Along the way, the treatise on scientific freedom would have to face the question of whether there ought to be "forbidden areas"—kinds of research so inimical to health or so morally degrading as to justify an outright ban.[27]

It would not place all the moral debts on the side of science. Society, it would argue, also owes science its cooperation and respect. On the list of blameworthy acts for which it would hold the other side accountable, it might include the government's interference with the free flow of scientific information through rules of clearance and classification, the covert employment of scientists by intelligence agencies, or the efforts by business interests to suppress unwelcome findings of research. The moral charting of scientific freedom, like that of academic freedom, would proceed from a mutual agenda, though the items therein would not be the same.

A treatise on scientific freedom would dismiss extramural freedom from the range of its immediate concerns. The profession-in-the-discipline has cause to protect the speaker as scientist, not as citizen; it can afford to leave the question of civil liberties to legal and constitutional solutions.[28] On the other hand, this treatise would repossess the principle of corporate freedom or autonomy, since the profession-in-the-discipline, through the medium of the professional society, has retained or acquired many guildlike features. What role the professional society should play in resolving conflicts over the hazardousness of research, or in pressing public agencies to abide by freedom-of-information laws, or in reproving members engaged in harmful enterprises, would be lively questions for theorists of scientific freedom; for theorists of academic freedom, they are not.

The authors of this document would seek to create a mechanism for investigating and reporting on suspected violations of their injunctions. In this respect, they would be following the course set by Committee A of the AAUP. But the monitors of scientific freedom, unlike the monitors of academic freedom, would not limit their patrols to campus precincts or deal primarily with threats to jobs. A charge that the standards for exposure to radiation set by the Atomic Energy Commission are too tolerant for human safety, a charge that data concerning the carcinogenicity of vinyl chloride had been doctored by scientists in the plastics industry, would be as much grist for this investigative mill as word that a scientist had been improperly dismissed.

The authors of this document would call for the creation of tribunals to resolve disputes over the acceptability of research. Doubtless they would look for pointers in the workings of academic tribunals, set up to prevent dismissals on ideological or other wrongful grounds. But they would soon discover that there is a world of difference between a pretermination hearing of a faculty member charged with professional unfitness and a nonadversarial examination of the dangers of a research project, and they would evolve a modus operandi that differed from academic due process in substantial ways.

In the end, the participants in this enterprise would come to realize that the wisdom needed to regulate science without destroying it has to be gradually acquired, and from many sources. That wisdom may be derived from the humane promptings of profession, the moral history of science, the arguments of libertarian philosophy, and the art of democratic compromise; it cannot be lifted bodily out of one venerable, but largely inapposite, norm.

References

[1]The qualifiers "conventionally" and "in this country" are meant to give contours to a subject that does not possess fixed bounds. Academic freedom, after all, is not a tangible creation, with uniform and universally accepted properties; it is at once a complex of ideas and thus subject to the vagaries of the imagination, and a congeries of values and thus subject to the variableness of human culture. One is at liberty therefore to define it as one pleases, and if one pleases to derive its meaning from flights of fancy, or from rummages in distant quarters of world history, no semantic court will ever issue a restraining order. But if one does do so, one will not obtain the most valuable meanings for instructional purposes. An idiosyncratic definition will be out of touch with the cumulative record of potent deeds that sustain academic freedom; an esoteric definition will not provide the social and institutional context in which academic freedom can best be understood. No claim is made that the definition treated here is more valid than any other. But it is more consequential and intelligible than any other, and it bids for priority of attention on these, if no other, grounds.

[2]On the putative connection between Germany's scientific eminence and her commitment to academic freedom, see Friedrich Paulsen, *The German University and University Study* (1906). A more searching and less encomiastic analysis can be found in Joseph Ben-David, *The Scientist's Role in Society* (1971). For a thoroughly iconoclastic view, see Fritz K. Ringer, *The Decline of the German Mandarins* (Cambridge, Mass.: Harvard University Press, 1969). The debt of American scholars to German training is well appraised in Jurgen Herbst, *The German Historical School in American Scholarship* (Ithaca, N.Y.: Cornell University Press, 1965).

[3]In Germany, academic autonomy acquired its special sacrosanctness from its medieval origins and from its subsequent harrowing career. During the sixteenth century, the universities of central Europe had lost their traditional independence when their church endowments were plundered and their papal protector was defied. From that time on, the confessional preference of the local prince had dictated the religious credo of the local studium, and a pall of sectarian exclusiveness had hung over the universities in the mottled territories between the Elbe and the Rhine.

There had been some loosening of constraint in the eighteenth century, when nonsectarian Halle and Göttingen opened their doors, and again in the early nineteenth century, when the University of Berlin, incarnating the idealism of patriot philosophers, was set up as citadel of free inquiry; still, even in the era of the Enlightenment, German professors had been expelled for irreligion, and even in the era of national revival, political purges of professors had occurred.

It was not until the Germanies were united under the Prussian Hohenzollerns that these universities could assert, with an air of confidence, that the period of church and state intrusion was forever past. Standing in happy contrast to the bigotry and possessiveness of the petty princes, the Wilhelmian policy of separation of church and state, and of generous support for academic science, lent credibility to this assertion. So, too, did the exercise by faculty bodies of some of the powers of their guild precursors, e.g., the power to set educational standards, to appoint assistants and instructors, and to elect their own presiding officers. Academic tenure, awarded to professors for life, barring serious offenses, gave further assurance that antique immunities had been restored.

It took an unusually critical eye to perceive that this Germany had not gone medieval, but had evolved an arrangement between faculty and ministerium tht left vital financial and appointive powers in the latter's hands. The powers to approve academic budgets, to fix the scale of faculty compensation, to set up and support new chairs, to veto faculty nominations for high positions all rested with the distant, not the resident, authority. Not a single Social Democrat was appointed to a full or associate professorship in the German Empire prior to World War I; this, and the disciplinary actions taken against radical privatdocenten appointed by the faculty, gave proof that the political interests of the state did not bow to the meritocratic interests of the corporation. Nevertheless, for the most part the German professoriate, perhaps because it was so deftly winnowed for loyalty and complacency, did believe that it enjoyed the essential attributes of the old consortio magistrorum. Cf. Friedrich Paulsen, *The German Universities, Their Character and Historical Development* (1895).

[4]Cf. Walter P. Metzger, "The German Contribution to the American Theory of Academic Freedom," AAUP *Bulletin*, 41 (Summer, 1955) 214–230.

[5]For a wide-ranging examination of the use of these concepts in American academic reform prior to World War I, see Laurence R. Veysey, *The Emergence of the American University* (1965).

[6]This is not to say that every codifier and guardian of academic freedom consistently adheres to this analytic distinction. Cf. American Civil Liberties Union, *Academic Freedom and Civil Liberties of Student in Colleges and Universities* (1st ed., 1949). Nor is this to say that the leading codifers and guardians in this field assert that persons in statu pupillari have no valid legal or moral claims to freedom of speech and association on the campus. Cf. American Association of University Professors, *Joint Statement on Rights and Freedoms of Students* (1967).

But the paramount concern of the former has been to advance and protect the rights of students as a matter of constitutional law, while the main thrust of the latter is to distinguish between student freedom as an educational desideratum (which it praises) and academic freedom as a functional necessity (which it defends). It is no accident that to this day, the AAUP has never considered an allegedly improper punishment of a student by academic authority, when unaccompanied by a similar offense against a faculty member, as grounds for a formal inquiry and report.

[7]That this was an inward-turning sociology, despite its references to society, becomes evident when one follows the trail of argument as it moves from the private to the public university. To refute the platitudes of private ownership to which certain trustees at that time were given, the authors of the 1915 report contended that the functional complexity of a private college or university—its involvement in the advancement of the sum of knowledge, its development of experts for public service—converted it from a creature of its founders into a kind of public utility. Just how many private institutions in this period were working on the frontiers of knowledge and contributing savants to public agencies, they did not say; doubtless, a census would have been discomfitting. But they were seeking a pro bono answer to the proprietary claims of all private boards, and for this purpose they were willing to equate the public significance of a Harvard or Johns Hopkins with that of the more common simple breeds.

When they turned to consider the state college and university, however, they found that the public service argument could be double-edged. Here, the regents, acting in the public's name, might exhibit the hubris of possession, and the public, demanding the services it paid for, might egg them on. At this point, the authors added a dose of Tocquevillian caution that subtly changed the initial argument. By "public", they indicated, we do not mean "public opinion", which can be tyrannous in a democracy; by "society" we do not mean the currently declared majority, which can be hasty and uninformed. To protect intellectual freedom in state-supported institutions, they tried to offset the claims of current taxpayers by invoking the interests of posterity, to hold the regents accountable to generations that were yet unborn.

The fiduciary argument, therefore, only appears to be extrospective, to start with the concrete needs of society and to work back to the university's obligations. In fact, it is highly introspective; it starts with the perceived presumptuousness of lay authority and then proceeds to rhetorical abstractions—first, the society-in-being, then the society-in-prospect—that serve as counterfoils. Cf. "General Report of the Committee on Academic Freedom and Academic Tenure," AAUP *Bulletin*, 1 (December, 1915): 21–24, 31–33.

[8]J. McKeen Cattell, *University Control* (1913).

[9]The reader must traverse the entire document before he becomes aware that the right of academic institutions to make their own decisions has not been defended as a component of academic freedom. The authors did express fear that "vested interests" and state legislatures might exert undue pressure upon the university, and they did argue for resistance by means of a vivid metaphor. A university, they wrote, "should be an intellectual experiment station, where new ideas may germinate and where their fruit, though still distasteful to the community as a whole, may be allowed to ripen until finally, perchance, it may become a part of the accepted intellectual food of the nation and of the world" ("General Report" *op. cit.*, p. 32).

At first glance, this rhetoric would seem not so much to disown the historic quest for academic autonomy as to seek out its contemporary equivalents. In place of the aggrandizements of Crown and Bishop, the imperium and sacerdotium in whose hands had lain the destiny of the studium, it appears to warn against the more up-to-date menaces of pressure groups and democratic legislatures; in place of the old-time guild, whose privileges and immunities had formed the traditional shield against external seizures, it appears to recommend an intellectual refuge, a sanctuary rather than a quasisovereignty, which the public may be prevailed upon to respect.

But a closer examination of the argument dispels the first impression of continuity. Phrased so generally and anonymously, these references to the world of might and power did nothing to expose the specific ways in which it encroached on academic institutions. Consequently, though these authors spoke with feeling about the need for a more sheltered university, they urged no structural reform to promote it—say, by securing for the state university a constitutionally independent status or by removing it from the petty oversight of state departments of education and finance. Instead, their recommendations for reform were confined to the internal operations of the university, and in particular those that affected employer prerogatives and employee rights.

[10]E. R. A. Seligman, "Draft Report", Committee A files for 1915, AAUP; Arthur O. Lovejoy to Seligman, September 27, 1915, Olin-Seligman file, AAUP.

[11]U. G. Weatherly, "Academic Freedom and Tenure of Office, Association of American Colleges *Bulletin*, 2 (April, 1916): 175–177; U. G. Weatherly to Seligman, October 26, 1915; James Q. Dealey to Seligman, October 26, 1915; Richard T. Ely to Seligman, n.d.; Charles E. Bennett to Seligman, October 31, 1915; Olin-Seligman file, AAUP.

[12]Cf. E. R. A. Seligman, "Committee of Academic Freedom" *Educational Review*, 50 (September, 1915): 184–88; Committee of Nine, *Preliminary Report of the Committee on Academic Freedom and Academic Tenure*, Pamphlets on University Education (December, 1914), pp. 1–8.

[13]"General Report," *op. cit.*, pp. 20, 37. In his article "Academic Freedom" written for the *Encyclopedia of the Social Sciences*, vol. 1 (1930) pp. 384–386, Lovejoy distinguished academic freedom from "political or personal freedom," in a retreat that is impossible to explain. However, he did say that a violation of the one would constitute a violation of the "spirit" of the other.

[14]"General Report," *op. cit.*, pp. 20, 24–25.

[15]AAUP, "Report of the Committee on Inquiry on Conditions at the University of Utah" (1915); "Report of the Committee on Inquiry Concerning Charges of Violation of Academic Freedom at the University of Colorado," AAUP *Bulletin*, 2 (April, 1916): 3–76; "Summary Report of the Committee on Academic Freedom and Academic Tenure on the Case of Professor Willard C. Fisher of Wesleyan University," AAUP *Bulletin*, 2 (April, 1916): 75–76; "Report of the Committee of Inquiry on the Case of Professor Scott Nearing of the University of Pennsylvania," AAUP *Bulletin*, 2 (May, 1916): 9–57; "Report of the Committee of Inquiry Concerning . . . the Dismissal of the President and Three Members of the Faculty at the University of Montana," AAUP *Bulletin*, 3 (May, 1917): 4–41. Cf. Walter P. Metzger, "The First Investigation," AAUP *Bulletin*, 47 (September, 1961): 206–210.

[16]Cf. the sharp judicial statement of this position by Oliver Wendell Holmes, Jr., when he was sitting on the Massachusetts Supreme Court. "The petitioner," he wrote apropos the complaint of a policeman dismissed by a local police department for venturing public criticism of it, "may have the constitutional right to talk politics, but he has no constitutional right to be a policeman . . . There are few employments for hire in which the servant does not agree to suspend his constitutional right of free speech, as well as of idleness, by the implied terms of his contract." McAuliffe v. Mayor of New Bedford, 155 Mass. 216, 220, 29 N. E. 517, 518 (1892). That the employment of teaching belonged to policeman's class was affirmed at the state level by the Tennessee Supreme Court in the famous Scopes monkey trial case: Scopes v. State, 154 Tenn., 105, 111–112, 289 S. W. 363, 364–365 (1927) and by the Supreme Court in Adler v. Board of Education of the City of New York, 342 U.S. 485, 493 (1952). Cf. William F. Murphy, "Academic Freedom—An Emerging Constitutional Right," *Law and Contemporary Problems*, 28 (Summer, 1963): 457–461.

[17]Cf. Walter P. Metzger, "Academic Tenure in America: A Historical Essay" in Commission on Academic Tenure in Higher Education, *Academic Tenure* (1973), pp. 128–135.

[18]Cf. William Van Alstyne, "The Specific Theory of Academic Freedom and the General Issue of Civil Liberty," in Edmund L. Pincoffs (ed.), *The Concept of Academic Freedom* (1972), chap. 5. See also Glenn Morrow, "Academic Freedom," *Encyclopedia of the Social Sciences*, vol. 1 (1968), pp. 4, 6.

[19]Cf. Walter P. Metzger, "Institutional Neutrality" in Carnegie Foundation for the Advancement of Teaching, *Neutrality or Partisanship*, Bulletin, Number 34 (1971), pp. 38–62. The general argument of the authors of the 1915 report was that a governing board could never legitimately fire a professor for his opinions, because a true university could never legitimately stand for any opinions at all. But it should be noted that they did make a concession on this score to academic institutions under religious auspices: "If a church or religious denomination establishes a college to be governed by a board of trustees, with the express understanding that the college will be used as an instrument of propaganda in the interests of the religious faith professed by the church or denomination creating it, the trustees have a right to demand that everything be subordinated to that end" ("General Report," p. 21).

On its face, this licensing of religious tests clashed with the neutral principle they attempted to enjoin. But these authors did not wholly take back the pardons and paroles that philosophy had wrested from theology, or all the territorial rights that secular scholarship had won in the several disciplines. While they did not declare against religious tests as such, they would not sanction the use of those tests in their most vexatious form. This is the inner meaning of their demand that religious requirements be open and above-board ("express understandings"). Tests imposed on faculty members subsequent to their appointments, and tests too vague to be understood by faculty members at any time, would not be approved, even under their permissive rule.

The purpose of these stipulations is clear enough: a doctrinal requirement imposed prior to an appointment acquaints the academic candidate with the character of the institution before he decides to cast his lot with it; the same requirement imposed after the appointment confronts the faculty member with the painful option of accepting a religious dogma he never bargained for or of surrendering a position he already holds.

Still, there remains the question of why the first generation of the AAUP, writing as it were on empty tablet, should accept the principle of religious tests in any form. They gave one reason in the text: very few important institutions of higher learning still retained sectarian attachments, and the trend in Protestant colleges was to move away from their religious past. It seems reasonable to suppose that they would not have been so yielding in this matter if Harvard had not long ago become desectarianized, or if Yale had not long since ceased to catechize candidates for positions in philosophy, or if the University of Chicago, with both a two-thirds Baptist board and a Baptist president, had ever thought it logical or desirable to appoint a Baptist faculty as well.

Beyond this lay an expediential reason for not declaring sectarian institutions out of bounds. In this country, the right of religious bodies to spread their beliefs and seek believers was secured by the early disestablishment of church and state, the constitutional guarantee of religious freedom, and the plethora of sects and churchs each claiming unique access to inspired truth. Protected by lenient laws of incorporation, easy rules of accreditation and a national climate favorable to private charity, these evangelical activities suffered little legal regulation or public oversight when they took on academic forms. An outright condemnation of religious tests would thus have clashed with the tradition of pluralism in religion and the tradition of laissez-faire in education, and would have been rashly enunciated by an association that was too young to have traditions of its own.

The trouble with this logic, as these authors fully understood, was that it had the power to overshoot its mark. If trustees with fixed religious creed were not required to become apostates and accept infidelity in the faculty, why should not trustees with fixed political and economic creeds enjoy the same exemption? If open-mindedness could not be thrust on a group with a definite vision of salvation, how could it be imposed on a group with a settled view of the good society?

It was Franklin Giddings, one of the members of the committee, who found a way to accept the religious exemption without hurling institutional neutrality, and thus academic freedom, over a fatal cliff. In his only contribution to the preliminary discussion, the noted Columbia sociologist suggested that the critical fact about a religious college was not that it was religious, but that it was "proprietary," that is, that it served the interests of its promoters rather than the interests of the public.

In this sense, it differed in no essential way from a college founded by a wealthy manufacturer to defend the social order, or from a school committed to Henry George's Single Tax. But Giddings did not fear that this synonymy would weaken academic freedom. For it could be maintained that all institutions of the owner type would have to accept the liabilities of its category—would have to announce that it had a proprietary purpose and thus forgo general public support. The sanctions that could not be imposed upon such institutions by a simple edict could be enforced by a professional insistence on honest labeling. Franklin H. Giddings to Seligman, November 1, 1915, Olin-Seligman File, AAUP.

[20]As yet, "no court has squarely held that there is a distinct right of academic freedom which elevates the status of the teacher above that of other public employees or citizens generally." "Academic Freedom," 81 Harvard Law Review 1065 (1968). Whether a teacher is sufficiently protected in his aprofessional utterances by the judicial interpretation of his first amendment rights is open to doubt. Cf. Pickering v. Board of Education, Will County Illinois 391 U.S. 563 (1968).

[21]"General Report," *op. cit.*, pp. 35–38.

[22]AAUP "The 1940 Statement of Principles on Academic Freedom and Tenure," *Policy Documents and Reports* (1973 ed.), p. 2.

[23]Carol S. Gruber, *Mars and Minerva: World War I and the Uses of the Higher Learning in America* (1975).

[24]It goes without saying that more occurs on this unheavenly earth than is contained in AAUP repositories. But there is no reason to believe that a survey of cases in other archives would reveal a markedly different pattern than that which the AAUP file discloses. The national Committee on Academic Freedom of the American Civil Liberties Union rarely gets wind of a professor being punished for work done in his study or his laboratory. The complainant files of the disciplinary associations may have more instances under this head, and they have not been consulted by this author. But it has been six decades since the leading associations in the social sciences, desiring to preserve their scholarly business from the distractions of scholarly complaints, turned over the burden of case inquiry to the broad-based professorial organization, and one may infer, from the ad hoc way in which most of these bodies now deal with the tenure complaints of their members, that they have not tried hard to retrieve the burden.

[25]Cf. Donald Light, "The Structure of the Academic Professions," *Sociology of Education*, 47 (1974): 2–28.

[26]A first and possibly important step in this direction was taken by the AAAS when its board of directors in 1974 charged its Committee on Scientific Freedom and Responsibility to study the problem of scientific freedom and responsibility and develop suitable criteria and procedures for its

further study. A report, prepared by John R. Edsall of Harvard University on behalf of this committee, was published in 1975. For a brief review of its major features, see John R. Edsall, "Scientific Freedom and Responsibility" *Science*, 188 (May, 1975): 687–693.

[27]In attempting to reconcile the competing arguments for freedom and responsibility, the authors of this document might choose to rely on the distinction rooted in law between freedom of expression and freedom of action, a distinction that would assign a more protected status to the enunciation of unpalatable opinions than to the conduct of dangerous experiments. However they might approach this knotty problem, they are not likely to follow the course of their academic counterparts who, seizing on the multiple clienteles of professors, conceded limits to freedom in certain functional areas in order to preserve a carte blanche somewhere else.

[28]Whether scientific freedom is itself a constitutionally protected right is another question. Current precedents and legal authorities would seem to hold that the right to scientific inquiry has no greater constitutional dignity than the conventional First Amendment freedom of speech and press. Cf. Harold P. Green, "The Boundaries of Scientific Freedom," a paper presented at the February 21, 1977, annual meeting of the American Association for the Advancement of Science.

SISSELA BOK

Freedom and Risk

FREE SCIENTIFIC INQUIRY and social stability are often at odds. Freedom of thought and inquiry threaten every form of authoritarian rigidity, every dogma by which men hold one another down. And when this freedom reaches into new areas—of acting upon people's minds, bodies, or environments, of using human beings as unwitting subjects, or of testing ever stronger means of destruction—a sense of alarm grows even in societies that have traditionally given free rein to research.

Communal self-defense and the protection of individual liberties then lead to demands for regulation of research with injurious or invasive potential. The freedom of scientists to pursue research unchecked must then be weighed against the freedom of those affected by the research. And the risk of hampering scientists and their work through regulation of scientific pursuits must, in turn, be weighed against the risk of harm in the absence of regulation.[1]

I would like to present some of the problems that arise in weighing such freedoms, such risks. These are *moral* problems: they concern choices by human beings of how to lead their lives and how to treat one another; of principles to guide such choices; and of standards of integrity, accountability, and concern. They arise in most efforts to seek knowledge experimentally: in the social sciences as well as in the natural sciences; in individual as in large-scale institutional or governmental experimentation; in basic as in applied research.

I propose to consider first the views that moral problems in science don't exist or don't matter or can be coped with by scientists themselves. I shall then illustrate the nature of the moral choices present in research, discuss the risks involved, and the burden of proof regarding these risks. Finally, I shall set forth three approaches to regulating scientific investigations according to the seriousness and complexity of the moral issues they raise.

Why Question Scientific Inquiry?

Many in the scientific community are disturbed that such questions should be raised at all. Their response takes two forms. "What risks?" ask some. Others wonder why, if there *are* risks, the traditional avenues of scientific self-regulation cannot suffice to cope with them.

The first question is often a genuine one. A great many scientists pursue tasks not even remotely capable of threatening anything or anyone. Many do observational studies, administer questionnaires, or work with purely statistical data. Others are engaged in basic scientific research from which only very indirect harmful effects, if any, can be imagined. To be sure, there is no clear dividing line between basic and applied research; too many research activities partake of both. Nor have risk and innocuousness ever fallen neatly on either side even of such an artificial barrier.[2] It is nevertheless safe to say that discernable direct risks arise more rarely in basic than in applied research. When scientists are engaged in inquiries they take to be harmless, they often generalize from their own activities, and may come to regard virtually all scientific investigations as fundamentally risk-free.

Others who do perceive some threats from certain kinds of research may consider the potential benefits to humanity as a sufficient compensation. Why, they ask, should our society not accept certain risks in order to conquer its most debilitating illnesses or shortages? Such a question is legitimate, but the *distribution* of these risks raises profound issues of a moral nature.

Can scientists, for instance, place specific persons at risk in order to benefit others? Can they create randomly distributed risks to some for the potential benefit of many? Does our generation have the right to place future generations at risk through our experimentation? Is is right for scientists from a nation banning certain experiments as too dangerous to carry them out in societies without such restrictions? However beneficial the final results that are hoped for, it is by no means self-evident that these results justify the experimental means employed. Regulation, even where the highest benefits are expected, cannot merely be waved aside without inquiry into the legitimacy of the assignment of risk, and especially into the question of consent by those placed at risk.

Whether scientists see risks from research as nonexistent or as merely distant and unimportant, the placid connotations of the very term "inquiry" fuels their puzzlement and their impatience with regulation. "Inquiry" in science evokes the solitary scientist thinking and questioning. Freedom of such inquiry is freedom of limitless thought and unfettered speech.

Steven Weinberg conveys a splendid image of the world of such a scientist:

> In the Science Museum in Kensington there is an old picture of the Octagon Room of the Greenwich Observatory, which seems to me beautifully to express the mood of science at its best: the room laid out in a cool, uncluttered, early eighteenth-century style, the few scientific instruments standing ready for use, clocks of various sorts ticking on the walls, and, from the many windows, filling the room, the clear light of day.[3]

These solitary, reflective, almost passive connotations of "scientific inquiry" do not in fact correspond, however, with many of the activities of today's scientists. These men and women are far from solitary in their interaction with others and in their use of vast public funds, far from passive in their use of powerful tools and machines and in the effect of their work on human welfare. They act upon nature and upon human beings in ways hitherto inconceivable. The freedom they ask to pursue their activities is, then, freedom of action, not merely freedom of thought and speech.

Even observational studies, in themselves seemingly least capable of having an effect of a harmful nature, can carry risks. These studies can hurt individuals through improper and intrusive observation, as where the subjects of research on private life are left in the dark about the fact that they are being observed. The means for such intrusive observations grow more ingenious each year; the research projects flourish.

Consider, for example, the social scientist who observed homosexual activities in men's rooms, then traced the participants, took notes on their homes and neighborhoods, and finally interviewed them without revealing his earlier observations upon them.[4] The harm from such studies can come from their intrusion alone; it can stem, also, from error or abuse of confidentiality in the communication of results.

Observational studies can also injure by taking the place of known therapy. When scientists who are also physicians merely observe patients to whom they have an obligation of assistance, their failure to act is unethical and in great need of regulation. Physician-scientists in the Tuskegee Syphilis Study "observed" the progress of syphilis in a group of patients without asking their consent to being in a study, or discussing with them possible forms of treatment. Even when penicillin became generally available, the study was continued. To have withheld such treatment without telling the patients is profoundly unethical no matter how noble the intention to learn more about the disease may have been.[5]

In questionnaires, too, inquiry can be improper. The questioning can be intrusive and bruising; the information gained can be misused and exploited. Political surveys, questions asked of the vulnerable and the powerless: these can turn inquiries into inquisitions. School children, for instance, are currently subjected to research on sexuality that can only be described as inexcusable prying. It is no accident that much research of a questionable nature has been conducted on the most vulnerable and helpless: on children, the institutionalized, the sick and the poor.

When it comes to more active forms of experimentation, finally, we are clearly even farther from the image of the solitary inquiring scientist. It is easy to see how ill the placid term "inquiry" fits much that seeks shelter under its umbrella. Research with dangerous drugs and ever more potent explosives can hardly be regarded as "inquiry" pure and simple. Freedom to conduct such research goes far beyond freedom of thought and speech. At the very least, such research demonstrates that the question "What risks?" is utterly inadequate to ward off all regulation of scientific inquiry.

The second question often raised by those wishing to protect scientific inquiry from the ravages of poorly planned or executed regulation is: if there is to be regulation, why not rely on *self-regulation*? Why should scientists not deal with abuses in research in the same way that they deal with plagiarism or the tampering with experimental results to expose them and render them harmless? Such internal policing seems the more necessary, the less likely it is that laymen will understand the complexities of modern research. Why not leave what regulation there has to be to those with long experience and the capacity to grasp what is at issue? George Ball, speaking about biological research, expressed strong support for such methods in his address to the 1977 annual meeting of the American Political Science Association:

> scientifically trained men and women are far better equipped to decide whether and how certain types of research should be conducted so as to safeguard to public interest than legislatures or administrative tribunals or courts. . . .
>
> [They] should be permitted maximum freedom to decide what research to undertake and how to undertake it, subject only to such safeguards as they might individually or collectively impose to prevent experiments being conducted in such a manner as to threaten the public health or welfare.[6]

Scientists argue, moreover, that the outside regulation which now looms—the committees, the commissions, the legislation, and the screening—will cut down both on the quality and on the quantity of research. Many, confronted by the bureaucracy and the paperwork already plaguing investigators, threaten to choose other lines of work. Freedom of inquiry is being stifled, they claim, because of nameless fears always directed to what is new and bold.

This discouragement is understandable. The growth of what has characteristically become an "ethics business" in our society is so spectacular that the temptations of zealotry, demagoguery, and illicit uses of power are strong. The original concern for ethics can then become exploitative and unethical in its own right. The bureaucracy of regulation of research can weigh as heavily as all other bureaucracies, and impede legitimate activity as much. Paradoxically, it can then allow genuine abuses to slip by unnoticed in the flood of paperwork required and minute rules to be followed.

Scientists see risks from regulation, therefore, that laymen often ignore—risks to themselves and to their capacity to work in their chosen fields. And the very passion with which opponents of research sometimes conduct their polemic—the conjuring up of diabolical visions of a future ravaged by our Faustlike cravings for forbidden power and knowledge—encourage equally dramatic and far-reaching counterarguments by defenders of research. They claim that the scientific undertaking as a whole is now in danger, that our capacity to survive is threatened once we shackle the exercise of curiosity, ingenuity, and reason. These capacities are brittle, they argue, and must be left to do their work in peace. Self-regulation by scientists is therefore in everyone's best interest; for its own sake, society ought not to try to interfere.

Such internal regulation is certainly indispensable in all professions. Yet it is far from sufficient for the protection of the public. Professionals have exhibited a pervasive inability to regulate themselves, whether in law, medicine, the military, or science. Severe abuses have been ignored; incompetents placing others at risk all too rarely expelled. In science, unethical research has seldom been combated by scientists themselves. The abuses of human subjects in biomedical research continued until the institution, in the last decade, of regulation. A recent UNESCO document on the status of scientific researchers, written in consultation with scientists, devotes much space to rights, but veils the subject of responsibilities in bland rhetoric.[7]

Self-regulation, then, does not suffice. What is needed is accountability not merely to colleagues, but to all who are at risk or their representatives. Neither the denial that research entails risk in the first place, nor the confidence in professional self-regulation obviate society's need to weigh freedom and risks for those conducting research as well as for those affected by it. The effort must be to refine, improve, and set just limits to the mechanisms of regulation, therefore, not to dismantle them.

Societies respond to the need for accountability in very different ways, if they respond at all; and different types of scientific inquiry are held accountable to different degrees. In the United States, biomedical research on human subjects has received increasingly strict inquiry and regulation.[8] Social science research, on the other hand, including large-scale social experimentation (on, for instance, education, housing, or income tax payments) is only beginning to receive outside attention of this kind.[9]

For all research, regulation is especially difficult whenever those at risk cannot be identified, or are very numerous. The difficulty is compounded when research is undertaken abroad—as when foreign scientists test contraceptive implants with unknown risks on women in Asia or Latin America, or when a nation tests nuclear devices far from its own shores and its major population centers. The risks may then be very differently understood and assessed on all sides; those who run the greatest risks may know least about them, or have insufficient power to impose regulation of any kind on the investigators.

The Freedom to Experiment

Freedom, risk, and benefit—these concepts have a long history of interaction in moral philosophy. How might we visualize their roles as we seek to evaluate scientific investigations? Are there, first of all, risky experiments where the freedom of the investigator should not be restricted?

Consider the research conducted by Anton Stoerck, a nineteen-year-old Viennese physician who wanted, in 1751, to demonstrate the safety of hemlock if used in the proper fashion, even though it was known as a poison since antiquity.[10] He had already applied hemlock solution externally, to his own skin and that of his patients, with no ill effects. He proceeded to drink increasingly large doses of the solution, once again with no mishaps. Only when he tried to taste the liquid extracted directly from the root of the hemlock plant did he experience great pain and swelling, fortunately temporary. He published these results, and claimed that further experiments showed that hemlock has "extraordinary virtue and efficacy . . . in the cure of Cancers, . . . Ulcers, and Cataracts."

Some of Stoerck's friends may well have thought him unwise to run such risks through experimenting on himself with a dangerous substance. Scientists might also think his experiment poorly designed for obtaining the knowledge he sought, and inadequate as a basis for the claims to curative powers. His method of preparing the hemlock solution may have been faulty; his dosage insufficient to prove his point. But it would be harder to argue that he should be *prevented* on moral grounds from performing such an experiment on himself, any more than that someone else should not risk mountain-climbing or childbirth.[11] It would be even more difficult to think of reasons why the government should regulate his autoexperimentation.

When, on the other hand, Stoerck went on to test his hemlock solution on his patients the moral aspects of his experiment altered drastically. At this point, it would have been perfectly proper to question the research on moral grounds. His freedom of inquiry could no longer be unlimited, as soon as he was placing other human beings at risk, no matter how important his aims, or how noble his intentions.

In such an experiment, it would have been appropriate (though unheard of on hospital wards until recent decades) to ask the investigator a number of questions before he could proceed: questions concerning what laboratory or other tests had preceded the use of human subjects; how well designed the experiment was for obtaining the desired knowledge (on the ground that *any* risk to others would be too great for an experiment doomed to failure); what risks the investigator foresaw for subjects and what safeguards he proposed in case of mishaps or untoward developments; what possible benefits to the patients themselves he envisaged from the experiment; how adequately he had informed the subjects of the purposes of the experiment, the procedures to be used, the risks attendant, alternative treatments available to them; and how free the subjects were to refuse to participate or to leave the experiment at any time.

Another way in which his experiment could have taken on moral dimensions is if he had undertaken it on himself alone but while he was responsible for the support of a family. Yet another moral concern would be present if he had acquired the funds to perform it from someone else; and a third, if he could foresee a direct and dangerous application of the knowledge he would gain. In each case, his research would affect others; in each, therefore, risks, alternatives, safeguards, advisability, and voluntary informed consent would matter to others besides himself.

Questions concerning these factors can be formulated and often answered. There is absolutely no reason why acknowledging that research has moral dimensions should lead to the familiar lapse into vague discourse about "values," followed by the conclusion that since such talk leads nowhere, the moral dimensions of research must, regrettably, be set aside.

In the discussion that follows, I shall concentrate on risks of relatively *direct* harm to human beings from scientific investigations. These may be risks of actual physical harm, of psychological harm, of coercion through force or deceit, or of invasion of privacy. They may stem from the process itself of carrying out the investigation; they may also stem directly from applying the knowledge gained, as with the working out of blueprints for cheap and simple nuclear devices that could be used by almost anyone with a grievance and access to fissionable material.

It is not always easy to know whether, and to what extent, research carries such direct risks. But the effort to see them clearly must be constant; for many invoke much more speculative, sometimes dogmatic reasons for restrictions on research. They may regard some forms of knowledge as threatening their own political or religious convictions. Perhaps they fear that the human spirit will be "stifled," or that societies will lose all spontaneity or adaptability in the face of new knowledge: the knowledge, for instance, allowing the average life span to be prolonged; or that of beings more intelligent than ourselves on other planets; or of any fact that might precipitate ideological changes.

Questions may arise about how and when to fund such research; individuals may or may not wish to participate in it. But to forbid research on the basis of such nebulous worries is not only unwise, but doubly illegitimate: it interferes with the liberty of investigators without adequate grounds; and it thereby interferes with the public's right to know. As Laurence Tribe has noted, "the Supreme Court has flatly rejected governmental measures 'whose justification rests . . . on the advantages of [people] being kept in ignorance.' "[12]

Nature of Risks

The *process* of carrying out scientific investigations can harm those, first, who participate: either the investigators themselves, their assistants, their human subjects, or bystanders. It can also harm persons at a greater distance, as when the process of research involves the spread of radiation or of toxic substances; at times it can also harm persons far from the site, as in the cancer deaths attributed to nuclear testing. The harm may be unforeseen, as when experimentation with the earth's atmosphere unleashes an unintended effect on the ozone layer, or on the contrary foreseen, as in the case of research with dangerous drugs in an effort to combat disease. Finally, the harmful effects directly attributable to an experiment sometimes only come to light long after the experiment is over. Unborn babies, for instance, may be exposed to drugs experimentally given to their mothers during pregnancy; but decades may elapse before the children manifest the symptoms caused by this exposure.

The *knowledge* that results directly from scientific experimentation can be equally dangerous. It can provide new, more easily accessible ways of harming people: blueprints for new weapons or for instruments to pry upon individuals and invade their privacy, or means for terrorizing entire populations. The dangerous knowledge may be a completely unexpected by-product of research, or it may be foreseen as a possibility but not intended, as when a drug developed to remove the symptoms of a disease can also be misused at high doses as a poison; or when confidential data in a study of, for instance, abortion or alcohol abuse come to public attention. Finally, the danger may be foreseen and intended, as in research on biological warfare and on ever more lethal weaponry. The magnitude of such research is staggering: out of an estimated worldwide expenditure in 1972 of $60 billion for research and development, $25 billion was for military purposes.[13]

Concern about possible harm from different experiments depends on the estimated probability that the harm will come about; this, in turn, depends on the degree to which it is a direct result of the process of research or the knowledge obtained. The more the intervening steps multiply, the more opportunities there are for forestalling or alleviating the harm. Concern varies also with the degree of severity of the possible harm, its extent, its irreversibility, and its capacity to spread. For this reason people respond very differently to a self-limiting risk of minor pain, on the one hand, and, on the other hand, a risk, however unlikely, of severe harm, rapidly spreading to ever greater numbers of persons, and impossible to arrest or reverse.

It is no wonder that individuals and groups have widely divergent views of what they regard as tolerable levels of these factors—tolerable kinds and probabilities of injury, for example, or tolerable severity or extent. The divergent views are affected by the degree to which those concerned are informed about the risks in the first place, by the political climate which allows or prohibits discussion of these risks, and by the sense of power or powerlessness to influence the conduct of risky research. These views are also very differently affected by the hopes of benefit from the research in question. Some may regard the risk of disease and even death for a small number as more than counterbalanced by the development of a way to combat a disease such as yellow fever. A few may even be willing to risk their lives in such an endeavor. Others may regard the matter with indifference so long as they are not themselves at risk, but refuse to share the danger.

A fundamental disagreement arises here with respect to consent. Some are willing to countenance a risky experiment so long as it is outweighed by expected benefits; others insist on the additional prerequisite that those at risk should have given their consent. In biomedical experimentation on human beings, such consent is now required by law. Other forms of experimentation give greater freedom to investigators, sometimes because the risks are deemed slight, sometimes when those at risk are hard to pinpoint or so numerous that they cannot all be asked to consent.

Some experimentation carries such flagrantly unwarranted risks or is conducted with such inadequate subject consent as to be clearly unethical in the eyes of all objective observers. This is the case when incompetent investigators handle dangerous materials or perform experimental surgery. In this category, also, falls research on prisoners whose consent has been obtained by coercion, and the experimentation by physicians in German concentration camps on such questions as the action of poisons or the length of time human beings can survive in cold water.

No one should have the freedom of subjecting others to such "inquiry" without the strictest procedures safeguarding the corresponding freedom of those at risk to know about the research and to refuse to participate. And there comes a level of risk to which not even consenting subjects or persons otherwise affected should be exposed. This level varies with the risks to these persons from alternative treatments. (Thus, a severely ill person may be willing to take chances with experimental surgery that someone with a milder form of illness would normally reject.)

Sometimes, on the contrary, the research is so innocuous that there can be no conceivable risk from it, either to any human subjects used or to others. Such processes as the gathering of data concerning the deceased, examining discarded fluids and organs of hospital patients in the laboratory, and a great deal of basic scientific research falls into this category. Other investigations may carry actual risks, but risks which are completely absorbed by the investigators themselves as in Stoerck's experiment.

But in much research, there is neither such clear abuse nor such obvious innocuousness. The risks, their magnitude and probability, may be unknown or disputed. The benefits hoped for are often just as conjectural. The information itself on which the choice should be based is in dispute or uncertain. In the controversy concerning fetal research, for instance, one empirical factor in dispute is at what stage of development a human fetus responds to pain. In laboratory research, the number of persons at risk may be disputed; when investigators claim they alone run all the risks in a particular experiment, their research assistants, others in the vicinity, perhaps the inhabitants of an entire region may argue that the burden is shared and that safety precautions should therefore be jointly chosen. And in a great many cases, the past experience from which predictions of safety or risk are extrapolated is itself difficult to specify.

Burden of Proof

Given the widely divergent estimates of risk from controversial studies, we must ask: should proponents of a controversial investigation have to demonstrate why their low estimates of risk are correct? Or should the burden of proof fall on those who oppose the research?

In much of the discussion concerning controversial research, the disputants take sides on this issue; but they rarely state an argument for their choice nor provide arguments against the opposite position. Thus, Carl Cohen, in discussing limitations on research with a high probability of very injurious consequences, states:

> Our rational commitment to freedom of inquiry is such that, in judging any claim of highly probable disaster, the burden of proof clearly rests upon those who would prohibit on that basis. They must present a convincing account of what concrete disasters are envisaged, what the methods are for determining the probability of such outcomes, and how those methods establish the high probability of the castastrophe pictured.[15]

If we used such high thresholds in other social decisions where there is some risk—regarding storm damage, for example, or war—we would take very few precautions indeed. Merely to avoid actions carrying a provable high probability of catastrophe is usually thought too careless an attitude for any society. Once again, the mesmerizing image of "scientific inquiry" has led some to adopt a double standard: they assign a recklessly high threshold of risk for action in the name of such inquiry but accept the normally more cautious threshold for nonexperimental action. They assume that scientists should have less of a burden of proof than others for the same level of risk.

Is this assumption reasonable? Should science be *exempt* to a greater extent than other undertakings from ordinary restrictions on action? Consider the exact same action—say, the drilling of holes in the crust of the earth—conducted by a scientific group seeking to learn how to control earthquakes, by a commercial establishment for financial gain, and by a government agency for defense purposes. Is there something special about the scientific purpose that lessens the burden of proof when the very same act is planned?

It would be unreasonable, I believe, to lessen the burden of proof for scientists alone at such times. Admittedly, scientific inquiry shares special protections when it comes to the freedom of thought and speech. But to the extent that scientific inquiry also involves action and direct risks, it has to be judged by standards common to other undertakings. If holes are to be drilled experimentally, therefore, we would expect those proposing such action to persuade us that the risks are worth taking—expect it as much from scientific investigators as from commercial or governmental entrepreneurs.

The burden of proof, then, rests on those proposing research carrying certain or possible risks. It rests, on the other hand, upon those wishing to interfere with research posing no apparent risk at all. And it can shift back and forth depending upon the cogency of the arguments offered in defense or in opposition of the proposed research.

In addition to asking where the burden of proof lies, we must also consider the degree of certainty of the proof of risk or harmlessness that is required. Even a catastrophe can sometimes be conclusively demonstrated only when it has already occurred. The same is true for assurances that nothing dangerous will take place. If total harmlessness were a prerequisite, little progress would be made in many areas where urgent needs must be met. Much will, therefore, depend on the standards of the decision-makers. An individual choosing whether or not to participate in experimentation as a subject can often allow himself to overlook or, on the contrary, to intensify the burden of proof to the point of lunacy; a committee or institution will have to adhere to standards that reasonable persons would regard as giving adequate protection.

Many despair of finding convincing solutions to these difficult problems of line-drawing, decision-making, and burden of proof. They may lapse into a facile relativism as a result, claiming that one solution is as good as the next, or that political clout will have to determine the outcome. Such a view can, however, be comfortably held only by those who will never themselves be the victims of an unjust policy; it does not make sense to hold it with respect to research policy, where each one of us may be among the victims should the wrong path be chosen.

We must, therefore, ask those who have taken sides on the issue of limiting controversial research to be very much clearer about how they formulate their arguments. We need to ask what they regard as a risk and as risk-free. We need to ask how carefully they have considered alternate forms of research leading to the knowledge sought. We need to know what kind of evidence underlies their estimates of risk; to what extent they rely on past experience, analogy, religious conviction; and where they locate the burden of proof. We need to ask what benefits they see as counterbalancing the risk, whose task they take it to be to weigh benefits and risk, and how informed consent will be sought from those whose decision it is.

Once the arguments are clearly set forth, it will at least be easier to determine where the disagreements lie, how cogent the arguments are, how acceptable the evidence, how suitable the decision-makers. The appeal of rhetoric and false analogies may diminish; and the substantive disagreements can be singled out for careful discussion.

Three Strategies

Three different strategies are needed in the face of moral problems posed by scientific investigations touching human lives. First of all, it is imperative that we become clear about the forms of research where no risks at all are posed to human beings, and that we work hard to remove the bureaucratic impediments from such innocuous research. (Decisions not to *fund* it may be made, of course, on many other grounds apart from moral ones.) Because of the growth of research activities and the confusion about what is and is not risky, there is now much needless harassment of investigators. It ought to be possible to set standards so clear that those who do harmless research can know from the outset that they will run into no roadblocks of an ethical nature.[16]

The second strategy addresses itself to the very opposite cases: those where all would agree that there is clear-cut abuse or recklessness. There is no need for complex moral reasoning to see that physicians should not deny treatment without consent to patients who suffer from curable illnesses merely in order to learn about the course of these illnesses; that social psychologists ought not to intrude surreptiously upon sexual or political activities; or that dangerous substances ought not to be tested in unsafe laboratories. There is no need to have lengthy committee meetings or earnest soul-searching about whether or not to allow such research. What is needed, rather, is the mobilization of public opinion and social change to combat such abuses perpetrated in the name of science. What is needed, here as for innocuous research, is to set clear standards, so that scientists can know beforehand when experimentation is too intrusive, or too dangerous to be undertaken. If these standards are clear and persuasive, they will in turn serve to reinforce individual standards of recoil by professionals from research that endangers or degrades human beings. Professional organizations may then come to take distance from such methods as sharply as they now condemn plagiarism or the falsification of data.

Experiments to be ruled out by such standards include the following:

—Experiments presenting some risk, yet so poorly designed that the results are worthless.[17] This common defect takes many forms, including inadequate sample, inadequate data-gathering, poorly formulated study aims, measurements not clearly defined, inconsistent research process from one instance to another, or inadequate links between expected findings and conclusions.

—Experiments presenting some risk performed by investigators lacking the requisite competence or skills.

—Experiments presenting some risk where the same knowledge can be obtained through comparable, but less dangerous studies.

—Experiments presenting some risk to human subjects whose informed consent has not been adequately ascertained. Deceptive experimentation, for instance, still pervasive in the social sciences, would thus be ruled out whenever it places subjects at risk.

—Experiments placing larger populations at substantial risk without thorough review and public accountability. Such public accountability is a surrogate for the individual consent that cannot be had from members of large groups.

These standards must be arrived at and enforced in the greatest possible openness. Government regulation cannot itself be relied upon to provide wise and adequate, yet not excessive, safeguards. In the first place, there are by now so many different regulations from so many government agencies that changing and conflicting rules provide the very opposite of simple, well-structured regulation. Second, in spite of all such rules certain government agencies have allowed, indeed initiated, some of the most sinister research, as well as some of the most careless experiments, quite incapable of reaching any results at all. Complete accountability and openness serve to limit such permissiveness; it allows the oversight of experimentation by representatives of those with most at stake, the individuals whose lives may be affected for better or for worse by the research.[18]

The third strategy is reserved for the more complex problems, and those where great disagreement persists. We must press the limits of the clearly intolerable and the clearly innocuous so as to make this middle group as small as possible, in order to avoid as much unnecessary dispute as we can. This narrower group of experiments must then be carefully discussed *before* we even reach the point of considering definitive social policy. The constant effort should be to sort out the easier-to-resolve problems from the harder ones; this effort should include, also, deciding which aspects of the hard problems are capable of quicker solution.

The hard problems that remain will require efforts at defining what is tolerable risk in a particular scientific investigation, when informed consent is adequate, when investigators are competent to undertake potentially dangerous research, and what forms of public accountability are sufficient for different levels of risk.

Time is a major factor here. The deliberation, sorting out, and looking for alternatives cannot be instantaneous. The public consultation and accountability take time too. Many difficulties with proposed research can be eliminated beforehand. Poorly designed projects can be improved; alternatives can often be found to reduce risk and to guarantee adequate consent. If, on the other hand, the difficulties only become apparent once the research is under way, there is much to be said for the process of the moratorium[19]: of ceasing forward progress for a time, while techniques are improved, safer alternatives worked out, risks evaluated, safeguards built up. But the costs of moratoria should not be forgotten: the loss of momentum, the discouragement of scientists who cannot easily shift gears; the postponement of possible benefits from the research. For these reasons, the moratorium is only appropriate as a last resort for the narrowed middle category. It ought never to be used as a weapon against politically unpopular research unless some demonstrable direct risk is at issue.

Time sufficient for deliberation and reasoned social choice is more important than ever when the damage feared is irreversible; and even more so when it may trigger irreversibly spreading harm. A moratorium can then allow an investigation of how reasonable the fears are, as well as the working out of ways to reverse possible damage and arrest its spread. It has the advantage of being reversible, unlike the damage itself.

The great obstacle to the taking of adequate time to consider risks and safeguards is the very natural individual drive to completion and achievement on the part of scientists, and the competition for priority, academic excellence, sometimes also for financial rewards and fame. The personal ambitions of individual scientists ought not, however, to be allowed to govern what is by now most often a joint undertaking, requiring joint safety and natural caution.

But in thinking about this third group of cases, there is no need to hesitate about the first two. Our society has every interest in speeding the efforts to eradicate clear-cut abuses in the name of research, and in lifting the many bureaucratic restrictions on innocuous scientific inquiry.

References

[1]For a discussion of the expressions "freedom to" and "freedom from," see Isaiah Berlin, *Two Concepts of Liberty* (Oxford: Clarendon Press, 1958); and Gerald C. MacCallum, Jr., "Negative and Positive Freedom," *The Philosophical Review*, 76 (1967): 312–334.

[2]See Hans Jonas, "Freedom of Scientific Inquiry and the Public Interest," *The Hastings Center Report*, 6 (August, 1976): 15–17.

[3]Steven Weinberg, "Reflections of a Working Scientist," *Daedalus*, 103 (Summer 1974): 45.

[4]Laud Humphreys, "Tearoom Trade — Impersonal Sex in Public Places," *Trans-Action*, 15 (January, 1970).

[5]See "Final Report of the Tuskegee Syphilis Study Ad Hoc Advisory Panel" (Washington, D.C.: U.S. Public Health Service, 1973).

[6]George W. Ball, "Biology and Politics," Address to the Annual Meeting of the American Political Science Association (September, 1977), pp. 13–14.

[7]UNESCO Recommendation on the Status of Research Scientists, adopted by the General Conference, November 20, 1974.

[8]See "Protection of Human Subjects," *Code of Federal Regulations* title 45 CFR part 46, November 6, 1975, and the Reports of the National Commission for the Protection of Human Subjects of Biomedical and Behavioral Research.

[9]See Alice M. Rivlin and P. Michael Timpane (eds.), *Ethical and Legal Issues of Social Experimentation* (Washington, D.C.: The Brookings Institution, 1975).

[10]See Anton Stoerck, *An Essay on the Medicinal Nature of Hemlock* (London: J. Nourse, 1760); and Lawrence K. Altman, "Auto-experimentation. An Unappreciated Tradition in Medical Science," *New England Journal of Medicine*, 286 (1972): 346–352.

[11]A perennial question in ethics is whether it covers not only what human beings do to one another, but what they do to themselves. "Duties to oneself" are discussed under this heading. Yet, whatever the resolution of this jurisdictional dispute, such duties are clearly of a very different nature for any discussion of outside limitation or regulation.

[12]Laurence H. Tribe, *American Constitutional Law* (Mineola, N.Y.: Foundation Press, 1978), pp. 904–905.

[13]*Disarmament and Development: Report of the Group of Experts on the Economic and Social Consequences of Disarmament*, (United Nations, 1972). See also Alva Myrdal, *The Game of Disarmament* (New York: Pantheon, 1976), pp. 11–14, 155–156.

[14]See Henry K. Beecher, "Ethics and Clinical Research," *The New England Journal of Medicine*, 274, (1966): 1354–1360; and Jay Katz, *Experimentation on Human Beings* (New York: Russell Sage Foundation, 1972).

[15]cf. p. 1207 in Carl Cohen, "When May Research Be Stopped?" *New England Journal of Medicine*, 296 (1977): 1203–1210.

[16]Periodic review could be required to ensure that unforeseen dangers, should they arise, do not go unnoticed.

[17]See David D. Rutstein, "The Ethical Design of Human Experiments," *Daedalus* (Spring 1969) 523–541.

[18]I discuss problems of accountability and risk in connection with deceptive experimentation in *Lying: Moral Choice in Public and Private Life* (New York: Pantheon Press, 1978).

[19]See Judith Swazey and Renée Fox, "The Clinical Moratorium: A Case Study of Mitral Value Surgery" in Paul A. Freund (ed.), *Experimentation with Human Subjects* (New York: George Braziller, 1970), pp. 315–357.

JUDITH P. SWAZEY

Protecting the "Animal of Necessity": Limits to Inquiry in Clinical Investigation

THE QUEST FOR KNOWLEDGE and improved technique always has been part of medicine. Clinical investigation and the role of physician-investigator, as a distinct field and occupation, however, are predominantly twentieth-century developments. In its brief history, clinical research has acquired a complex body of informal and formal controls, from within and without the medical profession, that has set "limits to inquiry" far more stringent than those encountered in the far longer history of research in the basic sciences.

As clinical research moves along a spectrum from laboratory research to clinical studies to accepted practice, the development of increasingly formalized and bureaucratic controls—such as peer review by institutional review boards (IRBs)—has had two major purposes.[1] First, as man increasingly has become the "animal of necessity" at certain phases in the study of human diseases and development of new preventive, diagnostic, and therapeutic modalities, controls have been instituted to protect human subjects from untoward risk, and, as expressed in the ethical and legal principles of informed consent, to promote and protect their individual autonomy. Second, the controls have sought to protect the public from the introduction and use of ineffective or unsafe drugs, devices, or procedures.

This paper will examine some of the formal controls over clinical research, explore the reasons they were instituted, and consider how well they meet their objectives. This inquiry, in turn, frames questions about the effects of various types of control over clinical investigation, and over scientific research more generally.

The increasing array of extraprofessional controls over clinical research, particularly the regulatory and legal ones, needs to be examined not only in terms of specific events that triggered their development, but also in relation to the types of professional social control that physicians in general, and clinical investigators in particular, can exercise informally over their own conduct and that of their colleagues. For, although the subject of professional controls generally suffers from a dearth of both empirical and theoretical content, what we do know about their nature and variable exercise by medical professionals helps to explain why extraprofessional controls have come into play.

One of the principal analysts of professional controls in medicine is sociologist Eliot Freidson. In nonpejorative sociological terms, he has described the physicians in a group practice that he studied as a "professional delinquent community" regarding their exercise of systematic self-regulation:

> The origin of the delinquent community of the physician lies in a position of vulnerable privilege rather than in one of vulnerable subordination. . .
>
> As a political and economic community of interest, the profession has been concerned with defending its privileged position by presenting a united front to the outside. . . . Given the profession's monopoly provided by the state, protected from the competition of all other kinds of claimed healers, the historic arrangement within the organized occupation was designed to sustain some modicum of collegial equality by which each colleague would be sufficiently satisfied so as to be inclined to close ranks without defection to the outside. . . .
>
> The process of social control in the medical group, I suggest, was rooted directly in the norms and customs of that historical tradition. In traditional . . . practice, there was no reason to discuss differences of opinion, knowledge, and practice with patients or colleagues. The market could solve interpersonal relations automatically, without the necessity of physicians undertaking deliberate effort at solution of conflict by reaching a consensus that could lead to changes in behavior and new forms of cooperation. . . . At best, [medicine's] delinquent community may be explained by reference to a sense of vulnerability to the possible imposition of . . . controls, an anticipatory response. The irony is that rather than defending the profession against the imposition of a rigid authoritarian framework, the anticipatory responses of the medical community have created part of the pressure for the institution of such a framework. The failure of the profession to control the availability, cost, and quality of the services of its members in the public interest—a failure tied directly to the internal laissez faire etiquette of the delinquent community—has contributed to the development in the United States of externally imposed requirements that may very well come to be, in the future, what the profession has always feared. It may thus have created what it has anticipated by the very defenses it erected![2]

Formal controls, originating from within as well as without clinical research, usually have been instituted in reaction to documented or alleged abuses of the rights and welfare of human subjects. When these abuses have been seen as signaling a weakness in or failure of internalized professional controls, they have been met with a variety of increasingly bureaucratic controls (often regulatory and legal).

The setting of limits to inquiry also needs to be appraised in terms of the diverse and growingly important range of influences exerted by "the public." Public or lay interest and involvement in various spheres of clinical research ranges from a seemingly insatiable mass media and literary market for medical research topics; to the focused activities of special interest groups concerning particular research topics and activities, such as psychosurgery, fetal research, and research with special populations such as prisoners and the mentally ill; to the role of community boards in approving, disapproving, and monitoring research in their areas.

Little empirical information exists about how and on what basis various sectors of the lay public assess clinical research and its effects. Nor, except in certain instances, has the direct influence of lay concerns on the generation of specific controls been documented.[3] But a number of projects violating basic ethical principles concerning experimentation with human subjects, and projects resulting in what is felt to be untoward harm, have received wide publicity in recent years. Public knowledge of such incidents has probably contributed to what many believe has been a profound shift in the public's valuation of clinical investigation. "There is today in this country," L. L. Jaffe wrote in 1969, "an enormous dynamic of human experimentation to which not only the medical profession but also the general public is heavily committed . . . an occasional catastrophe will touch off a public outcry with consequent overreaction, but on the whole the public continues to finance and to applaud medical experimentation and discovery."[4] How striking is the contrast between this statement and that written in 1977 by the editors of *Ethics in Medicine:* "From the idea of an ethical call to do clinical research, we have moved in less than a decade to medical research viewed in the public eye as a suspect activity that should require the prior approval of a governmentally constituted authority for each project."[5]

This perceived swing in public attitude seems to reflect a complex interaction between several features of the current public milieu, in addition to knowledge of particular suspect research activities. In recent decades, for example, science and medicine have been increasingly "demystified" and "secularized," and researchers often have been portrayed as competitive, political human beings engaged in work that is largely routine and often boring. Contrast the image of research conveyed to the layman in Sinclair Lewis's classic, *Arrowsmith*, with that conveyed in more recent works such as James Watson's *The Double Helix* and Joseph Hixon's *The Patchwork Mouse*. Science and medicine also have come to be perceived as big business, as complex, technologically sophisticated, and highly expensive enterprises, and one senses today the revival of a populism expressed in part as a conflict of interests between the little citizen and big business of science or medicine. Opinion polls indicate that laymen feel disenfranchised regarding policy decisions about science and technology, while seeing themselves as having a right somehow to influence those decisions. Some of these tensions in public attitudes and expectations deserve scrutiny. For example, it is striking to contrast calls for stricter bureaucratic controls over various aspects of clinical research in order to protect the autonomy and welfare of the public, on the one hand, with demands for individual freedom from regulatory authority to permit the use of substances ranging from saccharin to Laetrile, on the other hand. Finally, Americans for a number of reasons seem to have declining expectations about the imminency of major "breakthroughs" in medical research; correlatively, individuals may be less willing for themselves or others to be placed at risk for what they perceive as relatively small incremental benefits from clinical studies.

Ethical Codes

Since Antiquity the medical profession has formalized principles and rules of conduct for its members in prayers, oaths, and codes. Modern codes of medical ethics, however, are generally traced to nineteenth-century England, a period when, as I. Waddington suggests, physicians were moving away from the eighteenth-century patronage system toward a system of professional "colleague control" of their activities.[6] Works on medical ethics such as Thomas Percival's 1803 treatise were concerned principally with relationships between practitioners rather than between physicians and patients. This focus on intraprofessional relationships has been characterized by C. D. Leake as medical etiquette, rather than medical ethics which "should be concerned with the ultimate consequences of the conduct of physicians toward their individual patients and toward society as a whole."[7]

Without denying that physicians have long been ethically concerned about relationships with their patients, as reflected in portions of the Hippocratic Oath, Leake's distinction between etiquette and ethics is useful for assessing the extent to which formalized codes of ethical principles reflect issues of concern to the profession and to the larger society. Codes of both ethical and legal principles focusing on the rights and welfare of patients are a very recent development in the history of medicine, stemming from the arena of human experimentation. The first such code was the Nuremberg Code, which was formulated in a 1947 court opinion at the trial of twenty-three German physicians for "war crimes and crimes against humanity." In rejecting the defendants' claim that they had conducted ethical human experiments, the court held that only "certain types of medical experiments on human beings, when kept within reasonably well-defined bounds, conform to the ethics of the medical profession generally." The court then described those bounds in terms of ten "moral, ethical, and legal concepts" now known as the Nuremberg Code.[8]

The Nuremberg Code has served as the basis for subsequent international and national codes pertaining to clinical research in general and to particular topics such as organ transplantation. Perhaps the most widely cited post-Nuremberg Code is the World Medical Association's Declaration of Helsinki, drafted in 1961, published in 1964 and revised in 1975. The original Nuremberg principles have been expanded in several ways, reflecting both increased recognition of how complex and nuanced are the normative issues surrounding research with human subjects, and the institution of new controls such as the United States Public Health Service Guidelines.

Both the Nuremberg and Helsinki codes begin by stating the reasons for which clinical research should be conducted. The "protagonists" of human experimentation, wrote the Nuremberg judges, "justify their views on the basis that experiments yield results for the good of society that are unprovable by other methods or means of study." Helsinki "protagonists" affirmed that "it is essential that the results of laboratory experiments be applied to human beings to further scientific knowledge and to help suffering humanity." They further recognized two types of clinical research, therapeutic and nontherapeutic—a distinction that has become important in ethical, legal, and regulatory specifications of the conditions under which certain types of research may be conducted. First, there can be clinical or medical research combined with patient care, which can be carried out "only to the extent that [it] is justified by its therapeutic value for the patient." (Or, according to the 1975 revision, by its "*potential* diagnostic or therapeutic value . . .") Second, there can be nontherapeutic research, "the essential object of which is purely scientific and without therapeutic value to the person subjected to the research." In such research, however, the 1975 declaration adds that "the interest of science and society should never take precedence over considerations related to the well-being of the subject."

However essential for medical progress, the codes go on to state, clinical research ought not to proceed unless it meets a number of basic ethical principles. If adhered to, these principles will act as major professional controls demarcating the boundaries of ethically permissible clinical studies along the dimensions of the purpose, design, and conduct of research and the gaining of a human subject's informed consent.

Consent may be viewed as the core ethical and legal principle governing clinical research.[9] For, if the canons of informed, voluntary consent are met fully, it would be difficult if not impossible ethically to carry out a clinical research study that did not also meet these codes' principles about the purpose, design, and conduct of research. A subject's consent, the first principle of the Nuremberg code states, must have at least four characteristics: it must be competent, voluntary, informed, and understanding (or comprehending). The duty of conveying information about the proposed research, and of ascertaining the quality of consent rests with the investigator, according to the principle, and "It is a personal duty and responsibility which may not be delegated to another with impunity." Moreover, full disclosure of information is required:

> Before the acceptance of an affirmative decision by the experimental subject there should be made known to him the nature, duration, and purpose of the experiment; the method and means by which it is to be conducted; all inconveniences and hazards reasonably to be expected; and the effects upon his health or person which may possibly come from his participation in the experiment.

An exception to full disclosure, however, was allowed by the framers of the 1964 Helsinki Declaration, who left a "loophole" for investigators obtaining consent for therapeutic research: "*If at all possible, consistent with patient psychology*, the doctor should obtain the patient's freely given consent after the patient has been given a full explanation" [emphasis added]. The literature on the issue of disclosure of information to patients suggests that this provision reflected a widely held stance among physicians about the use of their professional judgment or "therapeutic privilege" in deciding how much information a patient (or in this case, patient-subject) *ought* to be given.[10] The 1975 Helsinki revision partially closed the loophole, reflecting requirements for institutional peer review of research protocols issued by the United States Public Health Service in 1966. It stated, "If the doctor considers it essential not to obtain informed consent, the specific reasons for this proposal should be stated in the experimental protocol for transmission to the independent [review] committee."

The sections of the Nuremberg Code and Helsinki declarations dealing with the design and conduct of research are concerned particularly with the difficult issues of risk-benefit assessment, in relation to the subject's well-being. "No experiment should be conducted," the Nuremberg Code states, "where there is an a priori reason to expect that death or disabling injury will occur."[11] Within these broad limits of risk, and in terms of the societally defined purpose of clinical research, the code further stipulates that "the degree of risk to be taken should never exceed that determined by the humanitarian importance of the problem to be solved," and that an experiment "should be so conducted as to avoid all unnecessary physical and mental suffering and injury."

The Helsinki declarations made some historically significant additions to these caveats. Reflecting the development of psychopharmacology and other behavior modification methods in recent decades, for example, the 1964 declaration states that "special caution should be exercised by the doctor in performing clinical research in which the personality of the subject is liable to be altered by drugs or experimental procedures."

Two new principles in the 1975 declaration reflect the development of peer review procedures since the mid-1960s, as well as concerns about the publication of scientifically and ethically questionable research studies. One requires that an experimental protocol "be transmitted to a specially appointed independent committee for consideration, comment and guidance." The second new principle asserts, "In publication of the results of his or her research, the doctor is obliged to preserve the accuracy of the results. Reports of experimentation not in accordance with the principles laid down in this Declaration should not be accepted for publication."

Legal Controls: The Law of Informed Consent to Human Experimentation

Ethical codes for clinical research are normative statements of the principles that *ought* to be followed in research with human subjects. Codes can operate only as informal controls, relying on the physician-investigator's personal and professional morality to implement principles about consent and the design and conduct of research. This fact is recognized in the 1975 Helsinki declaration, which stresses that, "the standards as drafted are only a guide to physicians all over the world. Doctors are not relieved from criminal, civil, and ethical responsibilities under the laws of their own countries."

In the United States and elsewhere, an extensive body of legal and regulatory controls has developed along with, and often drawing upon, ethical codes. In this and the next sections, I will briefly examine the limits placed on clinical research by three such sets of controls: the law of informed consent to human experimentation, and the regulations for clinical research promulgated by the Food and Drug Administration and the National Institutes of Health.

In their comprehensive review and assessment of the law of informed consent to human experimentation, Annas, Glantz, and Katz pointed out that the Nuremberg Code was the first, and remains the most, "comprehensive and definitive statement" of legal principles in this area.[12] Prior to the 1960s, indeed, little law had been developed in the United States or elsewhere "to regulate research organizations or investigators in their research methods, their areas of research, or the use of subjects or patients in that research," as W. J. Curran observed.[13]

Although "the law concerning human experimentation could be expected to develop on a case-by-case basis in traditional common law fashion," as Curran suggests, it was partly because of "the need to develop acceptable standards of care as an aid to the courts (admittedly in defense against arbitrary establishment of rules of conduct by the courts)," that the clinical research community in the late 1950s began both to study existing research practices and to draft codes or guidelines.[14]

When compared to the burgeoning malpractice litigation, and to the increasing number of federal regulations governing medical research, the number of court cases dealing with human experimentation remains small. Much of the law deals with issues of consent, the most central and the most controversial aspect of human experimentation. Investigators, subjects, and members of review committees know well how difficult and uncertain can be the interpretation and application of the ethical principles of informed consent to the myriad types of research with human subjects. Lawyers and judges too have confronted numerous and complex problems in determining the legal meaning of "competent, voluntary, informed, and understanding" consent. Developing a consent doctrine for the competent adult subject has been a slow and arduous legal task, and the difficulty of the task has increased as the law has attempted to deal with the meaning and exercise of consent for special subject-populations. For example, do the bonds inherent in the therapeutic relationship between patient and physician permit a legally valid consent to experimentation? Are minors competent to consent to participation in clinical research? Does the consent of a mother to research on a fetus provide sufficient protection for the interests of the fetus? Does the fact of incarceration prevent a prisoner from giving voluntary consent? Are all institutionalized mental patients precluded from giving competent and voluntary consent because of their diminished mental capacity and their incarceration?

The law's response to these questions is an evolving one. Some questions, such as the validity of a parent's or child's consent for the child to participate in nontherapeutic research, have yet to be decided as cases or statutes in the United States. Consent procedures and doctrine about the permissibility of research more generally have been shaped by decisions on specific issues such as transplants of organs from donors who are minors, and research with prisoners, institutionalized adult and minor mentally infirm patients, or fetuses, and with the competent adult participant in therapeutic research. These decisions show, in Jaffe's words, that common law judicial control of clinical research is marked by its flexibility, "a characteristic consonant with the presently fluid condition of ethical attitudes toward experimentation." In part because "resort to the courts is spasmodic" in the area of human experimentation, Jaffe goes on to observe,

> We cannot look to the common law judges for detailed prescriptions and proscriptions. Nevertheless, we must posit a common law as the ultimate legal guardian of the interests involved in experimentation. Where there is serious debate concerning the propriety or the necessity of certain procedures, where there is a real conflict of interests, an appeal to putatively relevant concepts of the common law provides authoritative standards for judgment.[15]

Clinical Drug Research: FDA Regulations

Of all bureaucratically mandated controls over clinical research, probably the most controversial in terms of their effectiveness and their effects are the Food and Drug Administration's regulations concerning the testing of new drugs. Beginning with passage of the first Pure Food and Drugs Act in 1906, legislation to control the development, manufacture, and use of drugs in the United States has been prompted, in significant measure, by strong public reactions to revelations about the injurious and/or inefficacious nature of various drugs.[16] Behind the provisions of the 1938 FDA law concerning proof of safety and labeling requirements, for example, lay the notorious case of the patent medicine Elixir Sulfanilimide, which caused the death of over one hundred persons from diethylene glycol poisoning.

It was not until the passage of the Kefauver-Harris amendments in 1962, following three years of congressional hearings, that the FDA explicitly was required to control the clinical testing of new drugs, and that manufacturers were required to establish the efficacy as well as safety of new substances.[17] Passage of the 1962 drug law might not have occurred without the public demand for stricter controls over the testing of new drugs following disclosure of deformities caused by Thalidomide in infants born in Western Europe during 1961 and 1962.[18]

The efficacy requirements of the 1962 law have had a pronounced effect both on the scope of clinical drug tests and on governmental regulation of those tests. The law requires three phases of the clinical testing of investigational new drugs, preceded by extensive animal testing, before approval for marketing can be considered by the FDA. In addition to specifying the types of studies that must be conducted to establish safety and efficacy, the regulations for all three phases contain other specific requirements such as submissions by the drug's sponsor to the FDA concerning the investigator's qualifications, the research design, progress reports during the trials, and consent requirements.

The consent provision of the 1962 law was added as an amendment in the wake of the Thalidomide tragedy because no state had a statute requiring physicians to inform patients that they were receiving an experimental drug. The final bill required that investigators certify to the manufacturer or other sponsor of a new drug that they will obtain the informed consent of human subjects; it had the same type of ambiguous exception, however, as the Helsinki declaration: "except where [investigators] deem it not feasible or, in their professional judgment, contrary to the best interests of such human beings."[19]

In 1966, addressing many problems of noncompliance with the new regulations and ambiguities of the consent provision, the FDA published its first interpretive rulings on consent in clinical drug trials.[20] Like the Helsinki declaration, the rulings required that consent always be obtained in nontherapeutic research and, seeking to remedy the ambiguities of the consent exception for therapeutic research in the 1962 law, the FDA tried to define more clearly the meaning of "not feasible" and "contrary to the best interests of such human beings." The FDA also added two important new consent requirements. First, subjects must be informed that they *may* receive a placebo or otherwise be used as a control in a clinical drug trial. Secondly, a requirement probably influenced by Henry Beecher's widely publicized 1966 paper documenting categories of "unethical or questionably ethical studies,"[21] a subject must be told of "the existence of alternative forms of therapy, if any."

The 1962 drug law and subsequent regulations provide, at least in theory, "the framework for excellent procedural safeguards for the protection of research subjects."[22] But like many other aspects of its activities,[23] the FDA's efforts to protect the subjects of clinical drug trials, and ultimately the wider patient population, have been faulted on many grounds. On the one hand, Lasagna, Wardell, and others, analyzing the development and introduction of new drugs in various countries, argue that the FDA's requirements for the avoidance of risk in the marketing of a new drug or a new use for an approved drug have become so complex, cumbersome, and costly that they have created a serious "drug lag" in the United States and have seriously hampered innovations in drug therapeutics.[24]

In counterpoint, there are concerns that, particularly as regards the protection of human subjects, the FDA's regulations and/or their implementation are inadequate. Legal experts, for example, have criticized on ethical and legal grounds the statutory exceptions to gaining a subject's informed consent. With respect to all such exceptions, Annas and colleagues hold that:

> Where the procedure is nontherapeutic, failure to disclose material risks would amount to fraud and should be condemned. In the experimental therapeutic setting the physician may have more influence over his patient, and the consent is not likely to be either voluntary or informed if material information is withheld.[25]

The General Accounting Office, in a report to the Congress, concluded more sweepingly that "federal control of new drug testing is not adequately protecting human test subjects and the public."[26] As of June 30, 1974, the GAO reported, the FDA was responsible for regulating some 1,400 sponsors and 9,400 clinical investigators, who were engaged in studying approximately 4,600 investigational new drugs in 250,000 human subjects. The FDA, with inadequate personnel and funds, has a staggering monitoring and evaluative task. Nonetheless, the GAO report charged, the "FDA has not aggressively regulated clinical investigations," and has made some efforts only in response to public pressures. The GAO reported,

> As early as 1968 FDA recognized that data submitted by clinical investigators was frequently unreliable and that on-site inspections of the facilities and records were necessary to insure data reliability. Not until June 1974, however, did FDA develop what it termed a "comprehensive plan for clinical investigation evaluation." According to FDA, the impetus behind developing the plan was "the lack of a concrete strategy . . . especially during a time of close public and congressional scrutiny resulting from the identification of unethical and non-scientific investigational studies." Although the plan appeared to be a step in the right direction, it had been only partially implemented.[27]

The GAO sent recommendations to the Secretary of Health, Education, and Welfare to strengthen various areas of the FDA's activities relating to clinical research. Concerns about the effects of over-regulation are valid ones. But so too is evidence that the FDA regulations are not ensuring ethically and legally tenable consent proceedings, and are not providing adequate checks on the design and conduct of drug trials themselves, prior to approving new drugs for marketing. Both sets of concerns, ultimately, must be assessed and responded to in terms of the FDA's complex primary mission, which is to regulate the industries that produce food, drugs, cosmetics, and medical devices, and in so doing to protect the consumers of these products.

The NIH Regulations

In fulfilling its mission to support a national program of basic and clinical biomedical research, the National Institutes of Health (NIH) has developed its policies and procedures for the protection of human subjects. The NIH and its parent agency, the United States Public Health Service (PHS), began to develop formal standards for human experimentation in 1953, with the opening of the NIH Clinical Center in Bethesda, Maryland.[28] In response to two concerns of the center's staff—how much risk should their normal volunteers and patient-subjects be exposed to, and how much information should they be given—the NIH soon issued its first set of principles and procedures for intramural clinical research.

Through the 1950s, the NIH advisory councils occasionally discussed the merits of applying guidelines also to extramural research, rather than, in keeping with the NIH's traditional policy of academic freedom, continuing to vest responsibility for the protection of research subjects with research institutions and principal investigators. Several events in the late 1950s and early 1960s raised the levels of concern within NIH about the use of human subjects in clinical research. In response to the interest and concerns evoked by the 1958 National Conference on the Legal Environment of Medical Science[29] and by Henry Beecher's "Experimentation in Man," a 1959 report which exposed ethical problems in human experimentation to the scrutiny of a larger medical and lay audience,[30] the NIH in 1960 awarded a three-year grant to Boston University's Law-Medicine Research Institute to study legal and ethical problems in clinical research.

The result of that study indicated that "in the medical research community, prior to 1962, there was a general skepticism toward the development of ethical guidelines, codes, or sets of procedures concerning the conduct of research."[31] Then came Thalidomide, the Kefauver hearings, and the new FDA drug law. Discussion of the need for guidelines governing extramural research with human subjects gained further momentum in 1964 due to the widely publicized Jewish Chronic Disease Hospital case, which involved the injection of live cancer cells in elderly patients without their knowledge.[32]

In 1965, the NIH sent recommendations to the United States surgeon general on the need to formulate principles about ethical issues in clinical research, and the need for the PHS-NIH peer review system to give special consideration to the use of human subjects. The National Advisory Health Council subsequently adopted a resolution stating that institutional peer review of research involving human subjects should be a prerequisite for Public Health Service funding, and in February, 1966, the surgeon general promulgated the first PHS Policy and Procedure Order for extramural research with human subjects.

> No . . . research or research training grant in support of clinical research and investigation involving human beings shall be awarded by the Public Health Service unless the grantee institution will provide prior review of the judgment of the principal investigator or program director by a committee of his institutional associates. This review should assure an independent determination: (1) of the rights and welfare of the individual or individuals involved, (2) of the appropriateness of the methods used to secure informed consent, and (3) of the risks and potential medical benefits of the investigation. A description of the committee of the associates who will provide the review shall be included in the application.[33]

The first of many revisions of the NIH Policies and Procedures for the Protection of Human Subjects was issued in July, 1966. Five years later the United States Department of Health, Education, and Welfare issued its *Institutional Guide to DHEW Policy on Protection of Human Subjects*. The *Guide* is considerably more detailed than the earlier PHS policy statement, and contains certain important additions, such as a fuller definition of informed consent and a broadening of the scope of subjects "at risk" to include possible psychological and social as well as physical harm caused by research. But it retains the basic clinical research control mechanism adopted by NIH, institutional peer review and approval of research proposals as a prerequisite for NIH funding.[34] Consonant with the freedoms and responsibilities that the NIH has assigned to its extramural grantees, the institutional review board (IRB) system primarily is a formal professional control mechanism. It relies on an investigator's peers, and secondarily on nonphysician members of IRBs, to examine and pass judgment on the risks and benefits of a proposed study and on the adequacy of the consent proceedings as described in the research protocol.

Like the FDA regulations, the NIH regulations have received mixed and often conflicting assessments. To some, the NIH controls represent "overregulating," presenting investigators with such time-consuming review requirements, and often trivial revisions of their protocols, that they are hindering the pace and course of clinical research.[35] To others, the peer review requirements of the NIH regulations offer necessary protection for human subjects, but there are often wide discrepancies between standards and practices.

In assessing the effectiveness of the NIH peer review regulations, as Bradford Gray points out, one needs to address four questions. (1) Do research institutions find the policy clear and workable? (2) Do institutions establish the required IRB mechanisms? (3) Do the IRBs have an effect on research proposals? (4) Do the IRBs affect the ethical and legal soundness of actual research?[36]

The key issues of whether and how the IRBs affect research proposals and the conduct of research itself have been addressed by three major empirical studies.[37] The results of all three studies are consistent along a number of dimensions, showing that while IRBs can and do have an impact on the design and conduct of research vis-à-vis the welfare of human subjects, there are numerous problems and deficiencies to be dealt with. These include problems with the composition of the review boards and relative influence of their members, their knowledgeability about the task of the IRB, and difficulties in interpreting various provisions of the NIH regulations. One such interpretive problem that many IRB members and investigators find particularly troublesome is whether their mandate includes assessment of the scientific merits of a research design, beyond a study's possible hazards. This is an uncertainty that seemingly should be resolved logically and clarified in the regulations if one reasons, as did the framers of the Nuremberg Code, that it is unethical to place subjects at risk if a study's design precludes obtaining valid and therefore useful data.

The studies also have documented numerous problems with the written consent forms now employed in a majority of clinical research projects. And more fundamentally, as emphasized by Bradford Gray: "in the absence of mechanisms to confirm that the goal of a procedure is being met, there is danger that the procedure can come to stand for the goal. . . . Thus, review committees must make clear that informed consent, not merely a signature on a consent form, is the operable goal."[38] This point is perhaps one of the most crucial issues concerning the effectiveness of IRBs vis-à-vis the actual conduct of research. For as studies have shown, although more systematic data are needed, there are investigators who choose to bypass required IRB review, and others who receive IRB approval and then alter their study design, even omitting consent proceedings. And even though investigators obtain written consent, their subjects still can be unwilling or uninformed participants. For, particularly among ward patients in teaching hospitals—a major population for clinical research—there are many forces that may induce patients to become subjects, often with little understanding of what that role entails.[39]

Although the NIH regulations have required a "continuous review" function of IRBs since the 1966 PHS policy statement, the nature of such monitoring remains unclear, and the need for it a matter of often heated debate. The impact of IRBs on the actual conduct of clinical research needs to be investigated more, but, as Gray concluded after his study, we do know that "research is being done that violates fundamental ethical standards." And, he adds,

> There seems to be no necessary reason why formal social control of human experimentation should rely primarily upon prior committee review. My own data show that while this procedure can have an important impact on proposed research, prior review is not effective in assuring ethical behavior in clinical investigation, particularly with regard to informed consent. If there exists a true commitment to assuring that research is conducted in an ethical fashion, a mechanism must be created that will detect unethical practices and make them visible.[40]

Limits to Inquiry and Professional Self-Regulation

As the development of the controls discussed in the foregoing sections illustrates, there has been a progressive shift in the locus of control from within clinical investigation to extraprofessionally or bureaucratically mandated laws and regulations. This shift, I have suggested, is due both to problems of systematic self-regulation within clinical research, and to public attitudes toward and expectations of medical research. One sees how these factors recurrently come together to produce a reaction to revealed abuses of the rights of human subjects, and to inadequate procedures for establishing the safety or efficacy of therapeutic innovations. To the examples we have cited, others may be added. These include the infection with hepatitis of institutionalized mentally retarded children at Willowbrook in New York, psychosurgery on incarcerated persons in Michigan and California, the withholding of treatment from subjects in the Tuskegee syphilis study, and, a recently revealed cause célèbre, the experiments with LSD and other behavior modification agents sponsored by the Central Intelligence Agency in the 1950s.

Most of the issues raised by these studies revolve about the role of informed, voluntary consent in the design and conduct of clinical research, as do the laws and regulations developed to address the issues. Given evidence that these control mechanisms have not been uniformly successful, new efforts to ensure the protection of human subjects continue to be mounted.

To this end, for example, the "National Commission for the Protection of Human Subjects of Biomedical and Behavioral Research" was established in 1974. The commission's creation was spurred in part by the findings of an ad hoc advisory panel convened in 1972 to examine the Tuskeegee Syphilis Study, a study of untreated black male syphilitics in Alabama, supported since 1932 by the Public Health Service. The panel, briefly, found that the study was ethically unjustified both because of its scientific design and because informed consent was not obtained from the hundreds of subjects. In its final report, the panel noted serious defects in HEW policies for the protection of human subjects. These included consent policies that were "poorly drafted and contain critical loopholes," "basic defects" in the existing IRB system, insufficient attention to the protection of special subject populations, and a lack of effective procedures for enforcing HEW policies when they were disregarded. To try to remedy such problems, the panel had recommended that Congress establish a "National Human Investigation Board," independent of HEW, "to formulate research policies, in much greater detail and with much more clarity than is presently the case.[41]

The panel's advice is reflected in the broad and powerful mandate given to the national commission. Its charge includes the study of two controversial research areas, psychosurgery and fetal research; the study of consent procedures for three special populations, children, prisoners, and the institutionalized mentally infirm; mechanisms for monitoring and evaluating the work of IRBs; "appropriate guidelines" for the selection of human subjects; and the "nature and definition of informed consent in various research settings." The commission is empowered to develop guidelines for biomedical and behavioral research, and make recommendations to the secretary of HEW for administrative actions necessary to apply such guidelines.[42]

A cluster of issues less visible publically relate to the design of research and to the analysis of results. In brief, these issues bear upon the protection not only of human subjects, but also of the larger public who ultimately receive new drugs, techniques, and procedures that presumptively are safe and efficacious. With growing evidence that many technologies often are not adequately assessed before they diffuse into widespread use,[43] there is growing attention by governmental and private groups to the need for mechanisms to better assure safety and efficacy, as was attempted for drug innovations in earlier laws and regulations.[44]

That there have been and are abuses of the rights and welfare of human subjects in clinical research is known, although the extent of the problem is uncertain. Whether the documented and publicized cases are relatively isolated, albeit particularly dramatic, episodes of unethical conduct, or whether they represent the tip of a far more pervasive problem, as is indicated by studies such as Beecher's, Barber's, and Gray's, needs far more empirical research, as do questions about the effects of existing laws and regulations.

Whether current perceptions about clinical research and discrepancies between standards and practices can be redressed, and if so by what mechanisms, clearly is a critical question for the future viability of clinical investigation, and, as it contributes to mounting questioning of the medical profession's "Aesculapian authority" in general,[45] for the viability of the medical profession as a whole. Clinicians engaged in organ transplantation speak of the "therapeutic tightrope" they must walk in using immunosuppressive drugs, attempting to tread a fine line that will prevent rejection of the donor organ without, at the same time, so weakening the patient's immunological defenses that the patient succumbs to massive infection. Similarly, in setting limits to inquiry in clinical investigation, we, as a society, must balance two important goals: to advance medical knowledge and technique through research, and to protect the rights and welfare of the human subjects who make that research possible.

The question is not whether there should be *any* limits to inquiry in clinical research, for the need for limits when man is the animal of necessity was recognized long before clinical investigation emerged as a field within medicine. The question, rather—a far more difficult one to address—is what those limits should be. What boundaries on the scope and conduct of clinical research will best equilibrate the goals of advancing medical capabilities and protecting human subjects? And, if we can define those boundaries, what control options will maintain them most effectively?

We are all directly or indirectly participants in the clinical research process, and a range of costs and benefits will accrue to us from the types of controls exercised over that process. The role and effects of ever more numerous, cumbersome, and restrictive bureaucratic controls in correcting the problems that seem to have eroded public confidence in clinical research need far more thoughtful scrutiny than they have received thus far in the catastrophe-reaction atmosphere that has generated most such controls. Possible policy actions demand, although they seldom receive, analysis of particular types of clinical research, definition of the controls extant or envisaged for those types, an understanding of how well each control is or is not working, and why, and assessment of the effects of the controls as they are presently exercised or anticipated to work.

Ultimately, I would argue, the difficult balance between advancing medical capabilities through research and protecting the subjects of that research can be maintained only if clinical investigators choose to exercise, individually and collectively, the array of informal and formal professional control measures at their command. As the experience with the FDA and NIH regulations suggests, externally mandated controls can proscribe the bounds of clinical investigation and can have a partial impact on some activities within those bounds, but they cannot, in the final analysis, guarantee ethical behavior in the design and conduct of research. Recently, this point was underscored when the State University of New York at Albany (SUNYA) was charged with "massive violations of federal and state regulations" for conducting internally funded psychology experiments without IRB approval and in some cases using allegedly uninformed and unconsenting subjects. R. Herdmann of the New York State Health Department, which brought the charges against the University, commented to *Science* magazine:

> What the incident at SUNYA demonstrates is that a substantial group of scientific researchers is unaware of society's interest in and concern for human research. These rules are not meant to be treated only as trivial paper exercises, and the problem is that when they are, there is a real chance of bureaucrats entering the university to do the monitoring—whether it be recombinant DNA research or

research on human beings. That would be a mistake, but it's a mistake that can happen if scientists cannot regulate themselves.[46]

REFERENCES

[1]In beginning to analyze the types, sources, interactions, and effects of controls, it has been helpful to classify them along several dimensions. The first dimension is what Freidson and others have termed "professional social controls" and "bureaucratic controls." In the field of clinical research, professional controls are those developed and exercised by researchers, and other physicians, in contrast to externally developed bureaucratic controls such as those required by an institution's administration, the courts, legislative acts, or regulatory statutes. As one looks at the ways that controls are exercised, however, it is apparent that they cannot be analyzed within these two categories alone. Thus, for example, there are bureaucratically mandated controls that are to be exercised primarily by physicians themselves, such as institutional review boards (IRBs) for research involving human subjects.

A second dimension is formality of operation of the controls. Informal controls operate principally through various kinds of informal or latent norms, largely dependent on values, attitudes, and interpersonal contact, and the consequences seen as flowing from the approval or disapproval generated in that contact. Many professional controls, including those developed through the physician's socialization experiences and his peer relationships, are of this informal nature. Formal control mechanisms, in contrast, as exemplified by most bureaucratic controls, are more structured and codified as to when and how they are used, and operate through more clearly defined lines of authority. Many controls have both informal and formal components. An investigator's decision about when it is appropriate to initiate a clinical trial, for example, can be affected informally by his own professional scientific and ethical judgments and by collaborative peer relationships, and by formal controls such as meeting the requirements of a funding agency, a regulatory agency, or an IRB.

[2]E. Freidson, *Doctoring Together. A Study of Professional Social Control* (New York: Elsevier, 1975), pp. 244–246.

[3]One such documented example of lay involvement in clinical research leading to new controls is Massachusetts' "Act Regulating the Use of Psychotropic Drugs in Children," passed in 1973. The law, which grew out of a 1972 controversy in Boston over a research project evaluating the effects of three psychotropic drugs on behavioral disorders in children, prohibits psychotropic drug research with school children in the state. See "MBD, Drug Research and the Schools," *Hastings Center Report*, Special Supplement (June, 1976).

[4]L. L. Jaffe, "Law as a System of Control," *Daedalus* (Spring 1969): 406.

[5]S. J. Reiser, A. J. Dyck, and W. J. Curran (eds.), *Ethics in Medicine* (Cambridge, Mass.: MIT Press, 1977), pp. 257–258.

[6]I. Waddington, "The Development of Medical Ethics—A Sociological Analysis," *Medical History*, 19 (1975): 36–51.

[7]C. D. Leake (ed.), *Percival's Medical Ethics* (Baltimore: Williams & Wilkins, 1927), p. 2.

[8]The Nuremberg Code and the Helsinki declarations are reprinted in numerous volumes; see, for example, Reiser, Dyck, and Curran, *op. cit.*, pp. 272, 328. For references to the Nuremberg Trials see G. J. Annas, L. H. Glantz, B. F. Katz, *Informed Consent to Human Experimentation* (Cambridge, Mass.: Ballinger, 1977), chap. 1.

[9]There is a large body of literature dealing with the principles and exercise of informed, voluntary consent. On the legal aspects of informed consent, see Annas, Glantz, and Katz, *Informed Consent, op. cit.*, A. M. Capron, "Informed Consent in Catastrophic Disease Research and Treatment," *University of Pennsylvania Law Review* 123 (December, 1974): 340–348. A sample of papers dealing with ethical issues in consent can be found in Reiser, Dyck, and Curran, *Ethics, op. cit.*, S. Gorovitz, et al. (eds.), *Moral Problems in Medicine* (Englewood Cliffs, N. J.: Prentice-Hall, 1976); and R. Hunt and J. Arras (eds.), *Ethical Issues in Medicine* (Palo Alto, Calif.: Mayfield Publishing Co., 1977).

[10]For a superb analysis of the ethical issues involved in disclosure, see R. M. Veatch, *Death, Dying, and the Biological Revolution* (New Haven, Conn.: Yale University Press, 1976), chap. 6.

[11]Recognizing the long and important history of physicians serving as their own research subjects, the code adds an interesting qualification: "except, perhaps, in those experiments where the experimental physicians also serve as subjects."

[12]Annas, Glantz, and Katz, *Informed Consent, op. cit.*, p. 1. While the Nuremberg Code has been used as an authority by only one United States court in a human experimentation case, the code

generally is recognized as part of international common law which can be applied in civil and criminal cases.

[13]W. J. Curran, "Governmental Regulation of the Use of Human Subjects in Medical Research: The Approach of Two Federal Agencies," *Daedalus* (Spring 1969): 542–543. Prior to 1947, court cases involved what is now classified as malpractice or quackery rather than clinical research, with courts using the word "experimentation" simply to mean a deviation from accepted medical practice.

[14]Curran, "Governmental Regulation," *op. cit.*, p. 545.

[15]L. L. Jaffe, "Law," *op. cit.*, pp. 408, 415.

[16]On the history of food and drug legislation in the United States, see J. H. Young, *The Medical Messiahs. A Social History of Health Quackery in Twentieth Century America* (Princeton, N. J.: Princeton University Press, 1967).

[17]Under the 1938 law, the FDA only required a drug manufacturer to obtain an investigational-use exemption, which permitted him to distribute a new drug interstate with a label "Caution—New Drug—Limited by federal law to investigational use."

[18]By the end of August, 1962, the FDA found that over 2 ½ million tablets of Thalidomide had been distributed to some 1,270 physicians; they in turn had prescribed the drug to about 20,000 patients. See FDA Press Release, August 23, 1962.

[19]Section 505(i), Federal Food, Drug, and Cosmetics Act.

[20]"Statement of Policy Concerning Consent for Use of Investigational New Drugs on Humans," *Federal Register*, 31 (August 30, 1966), p. 11,415.

[21]H. K. Beecher, "Ethics and Clinical Research," *New England Journal of Medicine*, 274 (June 16, 1966): 1354–1360.

[22]Curran, "Governmental Regulation," *op. cit.*, p. 555.

[23]J. Herzog, "Recurrent Criticisms: A History of Investigations of the FDA," Center for the Study of Drug Development, University of Rochester Medical Center PS-7702, (March, 1977); G. F. Roll, "Of Politics and Drug Regulation," *ibid*, PS-7701, (January, 1977).

[24]See, for example, S. A. Mitchell and E. A. Links (eds.), *Impact of Public Policy on Drug Innovation and Pricing* (Washington, D.C.: 1976); W. Wardell and L. Lasagna, *Regulation and Drug Development*. (Washington, D.C.: The American Enterprise Institute for Public Policy Research, The American University, 1975).

[25]Annas et al., *Informed Consent*, *op. cit.*, p. 54.

[26]Comptroller General's Report to the Congress, *Federal Control of New Drug Testing Is Not Adequately Protecting Human Test Subjects or the Public* (Washington, D.C.: General Accounting Office, July 15, 1976).

[27]*Ibid*, p. 20. The types of problems detailed by the GAO report are illustrated by its inspection of data from a 1972–1974 special survey of clinical investigators conducted by the FDA itself. Of 155 investigators inspected, 74 percent "failed to comply with one or more requirements of the law and regulations." These failures included failures to follow consent requirements, improper drug accounting, lack of adherence to the study protocol, inaccurate or unavailable records, and failure of the investigator to fulfill his or her required role in the research.

[28]Unless otherwise noted, material on the development of the NIH regulations is drawn from M. S. Frankel, *The Public Health Service Guidelines Governing Research Involving Human Subjects: An Analysis of the Policy-Making Process* (Washington, D.C.: George Washington University Program of Policy Studies in Science and Technology, 1972).

[29]"Clinical Research: Legal and Ethical Aspects, Sessions III," *Report of the National Conference on the Legal Environment of Medical Science* (National Society for Medical Research and the University of Chicago, 1960).

[30]H. K. Beecher, "Experimentation in Man," *Journal of the American Medical Association*, 169 (1959): 109–126.

[31]Curran, "Governmental Regulation," *op. cit.*, p. 548.

[32]J. Katz, *Experimentation with Human Beings* (New York: Russell Sage Foundation, 1972), chap. 1.

[33]Quoted in B. H. Gray, *Human Subjects in Medical Experimentation* (New York: Wiley Interscience, 1975), pp. 12–13.

[34]Most institutions now require compliance with the NIH regulations, including IRB review, for *all* research projects, whether or not they are funded by NIH.

[35]I know of no data to support the frequently voiced concern that the NIH regulations are causing a general "clinical research lag" in the United States. Many investigators, however, including this writer, can document the often inordinate length of time the IRB process can consume, to the point where the period of a proposed study may expire before IRB review and approval. In a recent national study of IRB procedures a substantial minority of those interviewed, particularly investigators, also felt that the review procedure "is an unwarranted intrusion on the investigator's autonomy, that the committee gets into inappropriate areas, [and] that it makes judgments it is not

qualified to make." See "Research Involving Human Subjects," A Report to the National Commission for the Protection of Human Subjects (Ann Arbor, Mich.: Survey Research Center, the University of Michigan, 1976), p. 5.

[36]B. H. Gray, "An Assessment of Institutional Review Committees in Human Experimentation," *Medical Care*, 13 (April, 1975): 318–328.

[37]B. Barber et al., *Research on Human Subjects: Problems of Social Control in Medical Experimentation* (New York: Russell Sage Foundation, 1973); Gray, *Human Subjects, op. cit.*, "Research Involving Human Subjects, *op. cit.* The study conducted by Bernard Barber and his associates was a 1969 questionnaire survey of some 300 clinical research institutions coupled with an intensive interviewing study of researchers at two institutions. The study by Bradford H. Gray examined in depth the workings of the IRB at a major university medical center, and the actual conduct of two clinical studies at that center. The third and largest study was conducted in 1974–1975 by the University of Michigan's Survey Research Center, for the National Commission for the Protection of Human Subjects. The study involved the review board activities and research at 61 institutions, including examination of protocols and IRB proceedings and interviews with a sample of some 3,900 investigators, review board members, and subjects.

[38]Gray, "Assessment," *op. cit.*, p. 376.

[39]Gray, *Human Subjects, op. cit.*, p. 242.

[40]*Ibid*, pp. 238, 250.

[41]"Final Report. Tuskegee Syphilis Study Ad Hoc Advisory Panel." (Washington, D.C.: U.S. Public Health Service, 1973).

[42]Public Law 93-348, sec. 202.

[43]Many drugs, devices, and procedures have been diffused into widespread use and later shown to be of limited or no efficacy; the history of surgery is replete with such discarded operations; others, such as internal mammary artery ligation, are still sometimes employed even though their usefulness has been discredited. Current technologies that have come into widespread use without adequate information concerning their efficacy or safety include the computed tomography (CT) scanner, mammography, and electronic fetal monitoring. In other cases, questions can be raised about the validity of studies that have been conducted to demonstrate efficacy or safety. For example, two studies of research reports in major medical journals found almost 75 percent of the publications that were analyzed to have unsupportable or invalid conclusions due to statistical problems alone. Other studies, which have looked at research design and data collection as well as data analysis, found that none of the reports analyzed had valid results; See A. R. Feinstein, "Clinical Biostatistics XXV. A Survey of Statistical Procedures in Leading Medical Journals," *Clinical Pharmacology and Therapeutics*, 15 (1974): 97; S. Shor and L. Karten, "Statistical Evaluation of Medical Manuscripts," *Journal of the American Medical Association*, 195 (1966): 1123; R. H. Gifford and A. R. Feinstein, "A Critique of Methodology in Studies of Anticoagulant Therapy for Acute Myocardial Infarction," *New England Journal of Medicine*, 280 (1969): 351; F. E. Karch and L. Lasagna, "Adverse Drug Reactions—A Critical Review," *Journal of the American Medical Association*, 234 (1975): 1236.

[44]One recent step was the passage of the *Medical Devices Amendments of 1976* to the *Food, Drug and Cosmetic Act*. The legislation provides for a new, comprehensive regulatory program for medical devices, centering on requirements for premarket clearance by the FDA. Groups engaged in studying intensively the need for better procedures to assure efficacy and safety include the congressional Office of Technology Assessment and, in the private sector, a Harvard University Interdisciplinary Seminar on Surgery; see J. P. Bunker, B. A. Barnes, F. Mosteller (eds.), *Cost, Risks, and Benefits of Surgery* (New York: Oxford University Press, 1977), especially Part II, "Surgical Innovation and its Evaluation."

[45]M. Siegler and H. Osmond, "Aesculapian Authority," *Hastings Center Studies*, 1 (1973): 41–52; see also the volume *Doing Better and Feeling Worse: Health in the United States*, *Daedalus* (Winter 1977).

[46]R. J. Smith, "Electroshock Experiment at Albany Violates Ethical Guidelines," *Science*, 198 (October, 28, 1977): 383–386.

BARBARA J. CULLITON

Science's Restive Public

THIS IS THE ERA of "public participation" in science. The once widespread feeling that scientists alone should have domain over the scientific enterprise is being replaced by a philosophy that calls for public involvement in science, irrespective of the fact that many of the elders of science find the idea abhorrent. The public has demanded a voice in decisions about the conduct and application of research and, to some moderate extent, that voice is being heard. In some instances, public involvement in science has more value in form than in substance. But in important ways, public influences are affecting the course of research itself. Certainly the advent of public participation has changed the social climate in which the scientific community works. There are things happening in the sciences now, in the mid-1970s, that were barely imaginable a decade ago.

At hospitals throughout the United States, nonphysicians now sit on the committees that review all protocols for human experimentation. The National Institutes of Health, in September, 1977, held a public meeting to evaluate and set policy for mammography, inviting laymen as well as professionals to present their views about the use of X-rays to screen women for breast cancer. A month later, the relatively unknown but congressionally mandated National Commission on Digestive Diseases began a series of hearings, to be held in nine states, "to hear from the people who pay the price [of digestive disease] in pain—the patients and their families." The National Science Foundation is struggling with a congressional directive to establish a "science for citizens" program to provide scientific expertise to citizens' groups that wish to challenge government policy on environmental issues, energy matters, or whatever. In at least three western states, decisions about the use of nuclear power have been decided by referenda.

And there is recombinant DNA—at present one of the more compelling issues. Questions about how, or even whether, to proceed with recombinant DNA experimentation have been asked in public for four years. In the early summer of 1976, the flamboyant mayor of Cambridge, Massachusetts, effectively and dramatically put recombinant DNA at the very center of debate about public participation in science when he threatened to prohibit that kind of research at Harvard and the Massachusetts Institute of Technology. Research with recombinant DNA has been on the agenda of official bodies in at least seven states in addition to Massachusetts. And at the top of the political hierarchy, bills that would govern recombinant DNA research are being considered by Congress.

If there is no mistaking that the era of public participation in science is upon us, neither can there be much doubt that the majority of scientists are still quite uncomfortable with the idea that anyone other than an investigator and his or her peers should have any voice in decisions about research. The scientists' spectrum of opinion runs from a grudging acceptance of public participation as a cumbersome annoyance to an alarmed view of the public participation movement as a menace to their right of free inquiry. As Massachusetts Senator Edward M. Kennedy said, ". . . academia has been on the defensive. It has chosen to view public scrutiny as a threat to scientific independence. It has chosen to view public involvement in particular research areas as inappropriate and representative of a trend toward antiintellectualism."[1]

Is the idea of public participation in science really a manifestation of anti-intellectualism? Does the public really want to limit scientific inquiry? What are the origins of the public participation movement and who is science's "public"? Just what does "public participation" mean? These are among the questions that require consideration in any attempt to understand the present social and political climate and its effects on the scientific community. Furthermore, they are pertinent to addressing a problem that was simply stated by Senator Kennedy and Senator Jacob K. Javits when they wrote in 1975: "Perhaps as the relationship between science and the public has evolved and changed in recent years, all parties have failed to understand both what is happening and what, ideally, ought to happen in the future."[2]

The idea that the public should somehow monitor and possibly curb science is a logical extension of the activist attitudes that became so powerful during the 1960s, coupled with a general realization that the application of new knowledge is not always to the good and with a deepening mistrust of the ability of established institutions to make good decisions. Large numbers of academics rallied in public protest over domestic social issues and, then, over the war in Vietnam. The environmental and consumer movements were growing. Welfare agencies were confronted by clients who demanded a say in the way programs for their benefit were developed. University students demanded, and in many cases won, what was said to be their right to sit on curriculum committees and to evaluate their teachers. It is not surprising that science too became an object of activist attention. People who feel left out of decisions that affect their lives have demanded that the establishment let them in.

University of Chicago philosopher Stephen Toulmin describes the present social environment with a striking metaphor: ". . . a lot of the difficulties that arise about the relations between the scientific community and the rest of society at the present time have important and significant parallels with the medieval problems of relations between church and state. . . . What we're faced with . . . is the Protestant Reformation. What we're faced with is a demand from the rest of society to be let in on the whole system of the ecclesiastical courts. . . . I think that [the public] suspect that the closure of the mechanisms of discussion is, in effect, a way of keeping them from debate about things that are really their business."[3]

But who is it that wants to be let in? The "public"? I think not; certainly not the proverbial man in the street. What evidence there is about him indicates that he has not lost faith in science, even though he realizes that the indiscriminate application of technology can be harmful. But there are no convincing data that the general public is antiscience or that it wishes to undermine the scientific tradition. What is happening is something else; the scientists themselves are challenging the way the research enterprise is run. Many of the voices calling for some change in ways of doing business are the voices of men and women who have made a serious, essentially full-time career representing what they describe as the public interest. Workers in environmental and consumer organizations thus constitute one of science's many publics. Some of those workers have training in science; most have some special interest in research or technology. All of them have invested a good deal of their professional lives and identities in being public representatives, but they surely cannot be equated with the man on the street.

Another important public is composed of practicing scientists. To a great extent, dissent against the traditional methods and presumptions of the scientific community is coming from within science itself. The difference is that the dissenters are taking their case to public forums. They appear in newspapers, on television, and before government bodies. Thus, when Harvard scientists who opposed certain kinds of recombinant DNA experiments failed to persuade the majority of their peers to take their side, they proceeded to take their fight, not to the university's board of overseers, but to the city mayor. One group of scientists has become, in effect, the public of another.

Science, of course, has other publics too, most of them falling into the category of special interest groups. In biomedical science, the American Cancer Society and the American Heart Association are a kind of public. Occasionally, the general public itself spontaneously makes its views about some scientific decision known. It did so in the spring and summer of 1977 when the Food and Drug Administration moved to take saccharin out of diet foods and colas. But this was a rare event, made possible by the occurrence of two circumstances. First, the issue directly touches the lives of millions of citizens who have no particular connection with science. Second, the public did not understand or fully accept the scientific data that convinced the FDA saccharin might be hazardous. And so when Congress voted to postpone the FDA's proposed ban, it did so in response to real public feeling. No discussion of public participation in science can ignore the pivotal role of Congress in making decisions that affect the course of science. Besides, it is the Congress that is so often the focal point of activities of all of science's other publics, for sooner or later each special interest public takes its case to Capitol Hill.

To acknowledge that for the most part the general public has little to do with science policy is in no way to diminish the role played by science's various publics. In this decade, two important series of events have occurred that will affect indelibly the course of biomedical research. One is the creation of the

National Commission for the Protection of Human Subjects of Biomedical and Behavioral Research, which grew out of a very public debate about experimentation on the human fetus. The other is the debate over research on recombinant DNA, an episode that has not yet concluded, but which is parallel in many aspects to the story of the National Commission, though the latter has to do with clinical research and the former a more "fundamental" science.

A comparison of the legislative origins of the National Commission and of proposed legislation governing recombinant DNA reveals the nature of the many publics of science and the manner in which they influence either specific research or the social and intellectual climate in which it takes place. It is often an adversarial process. In the case of recombinant DNA, public involvement depended totally upon the openness of the scientists who brought the issue before the world, and upon the opinions of scientific experts on both sides. In each case, public involvement and subsequent regulation occurred in areas that assumed public importance because of scientific progress. In neither case is there strong evidence that large numbers of citizens or legislators wished to limit free inquiry. Rather, they proceeded from a desire to protect individuals and society at large from harm. The intent is to control action, not thought. But it can be argued that the new sense of public awareness and the climate it engenders have put at end to the myth of the scientist-scholar free to follow his experimental life wherever it may lead.

Recombinant DNA research marks a maturing of biology. In the early summer of 1973, scientists attending the Gordon Conference on nucleic acids first fully realized that techniques for combining genes from one species with those of another—recombinant DNA techniques—were well developed. It meant that scientists could create in the laboratory brand new organisms that had never existed in nature. It was very heady stuff. But the scientists present at that conference were not so swept away with excitement that they failed to take clear notice of the fact that careless use or misuse of the new technology could be hazardous. They decided to go public with their concerns. Conference leaders Maxine Singer of the National Institutes of Health and Dieter Soll of Yale, acting on behalf of the conference, wrote to the presidents of the National Academy of Sciences and the Institute of Medicine. They called attention to the theoretical hazards of the experimentation and urged that the issue be carefully considered; they published their letter in *Science*.[4] The Academy convened a committee chaired by Paul Berg, a Stanford biologist who was a leader in the recombinant DNA field and a man convinced of the need to proceed with caution. Within a year, the Berg committee had met and come to the conclusions that an international conference should be held to assess the safety issues and, most startlingly, that in the interim, scientists should observe a voluntary moratorium on certain types of potentially troublesome recombinant DNA experiments. Again, the issue was put before the public. Berg and his colleagues called for a moratorium at a press conference held at the Academy in July, 1974, and formally made their appeal in a letter published that month in *Science* and *Nature*.[5]

These were extraordinary events, and in some quarters, Singer, Soll, Berg, and the others were roundly criticized for voicing their concerns in the first place and endorsing the idea of a moratorium in the second. But they took the steps they did out of a great sense of responsibility to the public, and not unmindful of the fact that the present revolution in biology might be analogous to the revolution in physics that gave scientists the knowledge to make an atomic bomb. The biologists chose not to remain silent. But they were hardly prepared for the public reaction that was to come. Their international conference, held at Asilomar in Pacific Grove, California, in February, 1975, drew wide and nearly universally laudatory press coverage. However, shortly thereafter, the voices of dissent were heard. The conference produced two pertinent recommendations, that the National Institutes of Health develop guidelines on recombinant DNA experimentation (the conferees produced a suggested draft), and that the moratorium on certain types of experiments continue until those guidelines could be developed.

At this point, the critics stepped in. Among their charges was that what was being proposed amounted to scientists asking the public to allow them to regulate themselves (which is precisely what they *were* asking) and that the public had been wrongly left out of the decision-making. This position was argued at a special congressional hearing in the spring of 1975 called by Senator Kennedy. It was arranged in the form of a debate with two scientists arguing on one side that the public needed to be brought into the act and two who had been active at Asilomar arguing on the other side—that only scientists could set effective policy because only they could understand the complexities of the science itself.[6] In addition, they testified against the possibility of federal regulation of recombinant DNA research on grounds that it would stifle freedom of inquiry. The circumstances of that hearing were typical of what was to come. It was a public hearing but there was no public participation (except for Kennedy). Scientists were talking to each other in public, competing for public approval, but the public itself did not have much to say.

As interest in the recombinant DNA debate began to grow, allegations that the public had been left out of the discussion were supplemented with claims that the research itself was an imminent hazard to the public health, not to mention a violation of the laws of God and evolution. And, as before, the dissenting voices were those of scientists (though not recombinant DNA researchers). The rhetoric escalated. Nobel laureate George Wald took the floor at a public hearing and likened an imaginary outbreak of recombinant DNA organisms to Legion fever—"unidentifiable and impossible to trace to its source." In a magazine piece entitled "New Strains of Life—or Death," Lieb Cavalieri of New York's Memorial Sloan Kettering Institute called forth images of a "new" Andromeda strain, neglecting to note that the old Andromeda strain is entirely fictional, and imagined a situation in which "cancer, normally not infectious, is spread in epidemic proportions by normally harmless bacteria."[7] Erwin Chargaff of Columbia University, in a letter to *Science*, referred to the "awesome irreversibility of what is being contemplated" by recombinant DNA experimentation, and said, "You can stop splitting the atom. You can stop visiting the moon; you can stop using aerosols; you may even decide not to kill entire populations by the use of a few bombs. But you cannot recall a new form of life. . . . An irreversible attack on the biosphere is something so unheard-of, so unthinkable to previous generations, that I could only wish that mine had not been guilty of it."[8]

On the other side, scientists talk about using recombinant DNA research to grow new types of plants to provide food enough for all the world's peoples and speculate that recombinant DNA studies may hold the cure for cancer. As to the matter of risks, Nobel laureate James Watson, at a New York State hearing on legislation, called recombinant DNA "the most overblown thing since . . . the fall-out shelter debacle."[9]

Recombinant DNA is an example of an issue in which the public and its elected representatives must depend upon scientists and their "expert" opinion in order to come to any sensible opinion about matters such as the relative risks as against benefits of the research. Indeed, as Cambridge mayor Alfred Vellucci noted in calling for city council hearings, he never would have taken on the issue had it not been for the fact that he had a group of illustrious scientists on his side. But sorting out the views of the two opposing sides is no easy matter. Faced with hyperbole about a subject that is difficult to understand in its scientific complexity, and confused by the spectacle of Nobel laureates and other leaders of the scientific establishment slugging it out in open meetings, legislators are turning to state and federal regulation—an imperfect solution to an insoluble problem—as might be expected. But legislators are, quite naturally, prone to legislate.

Recombinant DNA is an unfinished story. Persuaded that the public has been left out of decision-making and that the risks are real, Senator Kennedy introduced a bill that would have created a national commission to regulate recombinant DNA research from Washington. For many months, it looked as if that bill would win full Senate approval. However, forceful lobbying by the scientific establishment (which is still learning how to deal with Capitol Hill) at least temporarily forestalled the creation of such a regulatory commission and switched attention to a less restrictive measure pending before the House of Representatives. In this discussion, the final outcome is not important. What is pertinent is that the entire recombinant DNA episode has contributed to the emergence of the traditional scientific community as a "public" in its own right, a public now engaged in what psychiatrist Gerald Klerman describes as renegotiating a contract: ". . . there is underway a renegotiation of the implicit contract between the larger society and the biomedical research community. Indeed, there is a renegotiation of the contract between our society and all the professions—medicine, law, the military and education. In this renegotiation, there are new alliances and new communication patterns."[10]

Among the new alliances or institutions to emerge from this renegotiation, the National Commission for the Protection of Human Subjects of Biomedical and Behavioral Research is a unique manifestation of the idea that the public should participate in science, that researchers should not be left to regulate themselves. Although it is not a regulatory body per se, the National Commission—mandated by Congress and appointed by the Secretary of Health, Education, and Welfare—is in a position to make recommendations that in a general sense amount to regulations. In an important sense, its authority and power derive from the public it represents. All of its business is conducted in public. All of its recommendations about what can be allowed in human experimentation are public. And the HEW secretary is obliged to accept them or explain in public, in writing, why not. The National Commission is for clinical research analogous to what such a commission might be to basic research in recombinant DNA and, by extension, other areas of biology. It is an idea that was resolutely resisted at first as an inappropriate intrusion into an investigator's right to inquiry.

The origins of the National Commission can be traced back to 1967 when South African surgeon Christian Barnard transplanted a human heart into a grocer named Louis Washkansky. For eighteen days, Washkansky lived on borrowed time, with a borrowed heart. His name was on the front page of newspapers throughout the world. Heart transplants captured the public imagination. Surgically, they were a real tour de force. Emotionally, they were compelling. But they also brought to public attention questions about the definition of death, about surgeons playing God, about the appropriate uses of medical technology. Furthermore, they raised ethical questions about the fair distribution of a scarce resource, for fresh cadaver hearts were in short supply.

In 1969, Senator Walter F. Mondale proposed establishing a national commission to contemplate these perplexing new ethical issues, with an eye to federal intervention of some kind. However, there were strenuous objections on grounds that the medical research community should be free to regulate itself. It was also said that the government should not try to institutionalize anything as personal (some said irrelevant) as ethics. Mondale's proposal did not get very far at the time, but it became part of the public record, ready to be taken up again at a later time.

There were other circumstances in the late 1960s and early 1970s that led Congress inexorably toward the idea of overseeing research from Washington. A string of horror stories about scientists who apparently lacked ethical standards was unearthed by the press and investigated at congressional hearings. There was the Willowbrook scandal involving the exposure of young institutionalized children to hepatitis virus. There was the case of the cancer researcher who injected cancer cells into elderly patients dying of the disease. There was the Tuskegee syphilis experiment in which affected men were left untreated as controls even after penicillin became available. And there were startling, if somewhat overdramatized, reports of scientists perfusing the decapitated heads of fetuses in studies of blood flow. There were isolated incidents, to be sure, not representative of the vast majority of human experimentation. But they certainly seemed to be outrageous. A climate for congressional action was being created.

Another very important element in this brief history is the 1973 Supreme Court ruling legalizing abortion. During the preceding few years, a number of scientific advances, ranging from the perfection of techniques for amniocentesis to progress in the chromosomal and biochemical analysis of cells, had made fetal research attractive. Investigators were anxious to study the living fetus in the womb as well as fetal tissues. The abortion ruling, they believed, would make a variety of new studies possible. What was not anticipated, however, was the strength and political effectiveness of "right-to-life" groups whose opposition to abortion spilled over into opposition to all fetal research as well.

By early 1974 bills had been introduced in several state legislatures and in Congress that would have banned fetal research. It was an unprecedented case of legislative intervention in biomedical science that was directed not so much against research as it was against abortion. Two groups of public fought against the proposed bans. One was made up of scientists who wanted to do fetal research. Another consisted of organizations especially interested in certain diseases; they wanted fetal research to go forward because they believed it would lead to desired medical progress. Parents of children with genetic diseases, for example, can make a very persuasive case for fetal research.

The fetal research controversy—every bit as emotional as that over recombinant DNA—posed a substantial threat to scientific inquiry, but it also forged new alliances between legislators and scientists that might otherwise never have come about. In Massachusetts, for example, a state legislator who knew next to nothing about scientific research introduced a bill, at the behest of antiabortion constituents, that would have banned research on all fetuses, living or dead. As a result of that bill, scientists—many of them fetal researchers—who had never dealt with anyone in the statehouse, came out of their laboratories into the political arena. Months of intensive negotiations followed, with two valuable results. First, the objectionable bill was substantially modified. Second, the experience forged an alliance between legislators and scientists that led to the creation of a new committee, comprising scientists and laymen, to advise the legislature on proposed bills that would affect research.[11] The fetal research controversy and related issues about mass screening programs to detect genetic disease led to similar confrontations and, then, new alliances between legislators and scientists in other states as well. Maryland, for example, now has a public commission with authority over all of the state's genetic screening programs, its function being to protect the interests of the public and of science.

The battles that were being waged before state legislatures were also being pressed on Capitol Hill where the stakes seemed even higher. Amendments to ban all fetal research were before both the House and Senate which were also engaged in a debate over the wisdom of creating what became the National Commission for the Protection of Human Subjects of Biomedical and Behavioral Research—fetuses included. The Senate, Senator Kennedy in particular, was for it, the House against; the outcome was apparent. In the Senate, it was conservative James L. Buckley who introduced an anti-fetal-research amendment, tacking it on to Kennedy's bill to establish a federal ethics commission. Then, in a compromise move, Kennedy introduced a "perfecting amendment" that called for a moratorium on fetal research instead of a permanent ban. The proposed commission would be instructed to study the issue and the ban would become permanent when the two houses of Congress met in conference to resolve differences in their respective versions of the National Research Act of 1974. The bill was signed though virtually no one thought it was very good legislation. For the first time, Congress placed a moratorium on a certain kind of research, temporarily limiting inquiry.

The National Commission, whose members are meant to represent the public, was appointed through the most political of processes but emerged as a body that does, in fact, represent a wide range of views. Its eleven members include only five scientists, one of whom was known for his prolife position. It has women, white and black, lawyers, and ethicists, both conservative and liberal. If ever there were a body in a position to limit inquiry, this one is, for it is empowered to ask not just how specific types of research might proceed but whether it should be allowed at all. Thus, it could have recommended against fetal research altogether. Instead, out of a process of give and take, of people of divergent beliefs trying hard to understand one another's point of view, there came a compromise which permits fetal research to be conducted as long as carefully spelled out ethical guidelines are followed. It has taken, and will most likely continue to take, compromise positions on research on prisoners, on children, on psychosurgery and on other groups and issues. In its painstakingly achieved compromise, it is probably most effectively representing a very heterogeneous public.

There is every reason to believe that the current trend toward more public participation in science is going to continue. There is little reason to think that it will limit scientific inquiry, as has been feared. What experience there is to judge by indicates that the public can understand complex scientific and ethical problems and that it can arrive at reasoned opinions. The public committee of citizens appointed to review the recombinant DNA issue for the Cambridge City Council did not satisfy those who predicted it would try to stop the research; rather, it accepted by and large the recommendations of scientists who want the work to go on. Nonphysicians who are members of institutional review boards have not brought a halt to human experimentation. Neither has the National Commission.

Public participation is not dangerous for the scientific enterprise. It is time-consuming, there is no doubt about that, as those who have defended clinical and basic research can well attest. And it is likely to lead to restraints that previously were not imposed. Nevertheless, the restraints that come from ethical considerations and recognition of the need for public accountability cannot be dismissed as inappropriate. To the contrary, they may lead in the end to greater public understanding of and appreciation for science. In any case, they are part of the social cost of democracy.

References

[1]Edward M. Kennedy, in an address at the Harvard School of Public Health, May, 1975.

[2]Edward M. Kennedy and Jacob K. Javits, in a letter to Willard Gaylin, president of the Institute of Society, Ethics and the Life Sciences, October 19, 1975.

[3]Stephen Toulmin, in remarks made at a conference on "Biomedical Research and the Public," Airlie House, April 1–3, 1976.

[4]*Science*, 181 (September 21, 1973): 113.

[5]*Science*, 185 (July 26, 1974): 303; and *Nature* 250 (July 19, 1974): 175.

[6]Hearing before the Senate Subcommittee on Health, April 22, 1975. For a report, see Barbara J. Culliton, *Science*, 188 (June 20, 1975): 1187–1189.

[7]Lieb Cavalieri, *The New York Times Magazine*, August 22, 1976.
[8]*Science*, 192 (June 4, 1976): 138–139.
[9]Hearing before New York State Attorney General, New York City, October 21, 1976.
[10]Gerald Klerman, in remarks made at a conference on "Biomedical Research and the Public," Airlie House, April 1–3, 1976.
[11]Barbara J. Culliton, *Science*, 187 (January 24, 1975) 237–241.

PETER BARTON HUTT

Public Criticism of Health Science Policy

PUBLIC CRITICISM OF SCIENCE and consequent attempts by the government to impose control over scientific inquiry are not recent phenomena. It is doubtful that scientific inquiry has ever been truly unfettered, or that any responsible scientist would argue for such anarchy.

It is likely, indeed, that scientists today enjoy greater freedom of inquiry than ever before in history, in terms of both public support through use of tax funds and reasonableness of government restriction.

Public criticism of current health science policy has not sprung full-blown upon an unsuspecting scientific enterprise. It began years ago, and has steadily increased in intensity. Perhaps because of its extraordinary success in attracting sustained congressional support for health science research during the past few decades, the scientific community has paid little attention to this growing criticism. Even when given the opportunity or mandate to analyze and rebut this criticism—for example, when Congress established the President's Biomedical Research Panel to review and assess national biomedical and behavioral research conducted and supported under federal funds—the scientific community has not dealt with it directly, and in some instances has ignored it completely.

Some of the criticism amounts, of course, to little more than rhetoric and hyperbole. Scoffing at even this type of criticism, without patient and objective analysis and a mustering of all available information to refute it, is a very dangerous mistake. Even the most absurd accusations, when repeated time and again without refutation, can gather a following and momentum that defy common sense.

Lawyers are trained to pay little or no attention to those who agree with them, and instead to concentrate on answering the arguments of their adversaries. The scientific community has regrettably given the impression that there are in fact no convincing answers to be given, because it has failed to meet the criticism of current health science policy head on, to debate it publicly, to concede those points where criticism is valid, to present a sound rebuttal where criticism is demonstrably unwarranted, to admit where further inquiry and information are needed to determine the validity of criticism and then to set up methods of gathering the needed information, and to pursue these issues in every public forum available.

In the scientific arena, the scientist is accustomed to setting up hypotheses, testing them with experimentation, declining to offer conjecture until adequate data are available to support or refute the hypotheses, and even then limiting his observations to those things that are objectively proved and waiting for new hypotheses and investigations before again venturing to suggest additional conclusions. In the public policy arena, in contrast, he is reluctant to engage in any disputation and indeed feels extremely uncomfortable about participating in any debate at all, he remains silent or relies upon conjecture and appeal to emotions rather than attempts at objective analysis, and he is affronted and bewildered when an uncomprehending public rejects what he believes to be self-evident propositions of sound science policy.

Scientists, if they are to be credible and, more important, if they are to be effective in maintaining the extraordinary level of freedom of scientific inquiry that exists today, must begin to apply more rigorously in the public policy arena the same principles of investigation and proof that have worked so well in the field of science itself. The scientific community cannot afford to permit criticism to continue unrebutted until it is accepted by the general public as undisputed statements of fact.

The first step in the process of rebuttal is to catalog the public criticism of health science policy in this country so that it can be analyzed and subjected to proof or refutation. The remainder of this paper constitutes an attempt simply to do this. The catalog is culled from a wide variety of sources, including hearings and reports of Congress, reports of government agencies, and speeches and position papers prepared by individual scientists and nonscientists and published or summarized in journals, magazines, and newspapers. It is not arranged in any order of priority, and is undoubtedly incomplete. However, it contains sufficient challenge to occupy the scientific community for some time to come.

Some of the criticism may seem totally unjustified and even purposely offensive. No attempt has been made either to embellish existing criticism or to tone it down; it is simply set forth in the terms in which it is often heard in public policy debate. Many of these issues are emotionally charged, and virtually all are advanced by supporters who believe fervently in the logic and righteousness of their position. The scientific community cannot afford to take offense regardless how it may resent the tone and innuendos, or even the nasty accusations, that some of this criticism entails.

Finally, it is important to emphasize that this catalog does not represent my own views or those of any other particular individual despite the direct declarative mood of the sentences in which it is written. It is intended to present a list of public criticism, not to debate the merits of the issues involved.

I. Health Science Research Is Not Presently Designed to Help the Public Which Pays for It

A. *Appropriations for health science research funds must be justified in terms of demonstrable and immediate public health benefits rather than in terms of vague promises and hopes that, in time, basic research will provide solutions to all health problems.* The public has been deluded too long with promises that all human disease can be prevented, controlled, or cured, if only more and more public funds are poured into health science research. Tax money is becoming scarcer and scarcer each year, and only those research projects that promise a tangible public benefit in the very near future should be regarded as worthy of federal funding.

B. *Little research is done to develop a national research policy attuned to public needs.* There have been only a very few published papers analyzing how health science research priorities should be established. Significant federal funding should be devoted to determining an appropriate strategy for setting such priorities, communicating them to the research community, encouraging research into important areas of national need, coordinating research on questions of major public significance, and assuring rapid technology transfer so that the public will receive maximum benefit.

C. *There is no attempt to set research priorities on the basis of the projected benefit to the public of reducing the public burden of illness.* The objective of federal funding of health science research should be to eliminate or alleviate the morbidity and mortality associated with those health problems that have the greatest impact on the public. The federal funding agencies should establish a method of pinpointing those health problems which impose the greatest burden on the public, determine the research priorities which have the greatest potential for affecting those problems, and then fund those research proposals which directly address those priorities. Scientists would remain free to pursue whatever other interests they may have using nonfederal funds, but scientific information that has no potential impact on public health is of little value to the public and should not be funded by federal agencies.

D. *The free enterprise market mechanism represents the best model for assuring that the right kind of health science research is undertaken.* It is in the best interests of private industry to determine where the greatest burden of illness falls on the public and to devote research expenditures to those areas, since they represent the greatest potential market and thus the greatest potential source for profit. This is the classic example of the free market working in the best interest of the public. Federal funding should either be subjected to the same principles or should be discontinued.

E. *Research priorities should not be determined by the interests of researchers.* Under present federal policy, research grants are awarded on the basis of research proposals submitted on the initiative of the individual researcher. Accordingly, it is the interests and priorities of researchers which largely control the nature of health science research undertaken through federal funding.

Those health science issues that are of primary importance to the nation, rather than those that are of primary interest to researchers, should receive targeted federal funding. Federal agencies should determine the most important health needs facing the country and then fund only those projects which are directed towards meeting those needs. Allowing researchers to follow their own interests is far less likely to bring results that truly reflect national health priorities.

F. *Research projects should not be directed to abstract and artificial issues that bear little or no relation to possible health benefits to the public.* Without doubt, there are many possible research topics that could be of intellectual and academic interest. But federal funds, which will be increasingly scarce in the future, should be used to support only those health science research projects that have the greatest potential for improving the public health.

G. *Research priorities should not be allowed to be influenced by the need a researcher feels to pursue the "latest" issues in order to keep up with his peers.* It is true that every individual is subject to peer group pressure, and that everyone wants to feel that he is part of the most interesting new developments in his chosen occupation. That is no excuse, however, for using federal funds to accommodate a researcher's own personal ambitions. Federal funds should be used only for the most promising health research needs, and only where there is not already an adequate number of qualified researchers pursuing the matter.

H. *Research priorities should not be allowed to be influenced by traditional academic views about important research issues.* Some scientific fields are regarded by the academic community as important intellectual challenges, and others are regarded as the backwaters of scientific inquiry. Academic leaders routinely advise their brightest protégés on the areas to pursue and those to avoid, for a successful academic career. The academic view of the relative significance of research subjects often bears little or no relationship to the actual health benefits that might be obtained from them. The policies that guide federal funding of health science research should be determined solely by the potential for public benefit rather than intellectual stimulation and academic honor. Targeted funding will do much to assure that barriers to important research on neglected areas are overcome.

I. *The academic emphasis on "publish or perish" leads a researcher to waste great amounts of time on trivial papers designed primarily to lengthen his curriculum vitae.* Anyone who has ever read the lengthy list of publications that routinely accompanies a research scientist's curriculum vitae inevitably wonders how many of those papers have real significance, and how many were pursued in order to assure academic tenure. The honest scientist will admit that his important and productive papers are few in number.

The proliferation of scientific publications threatens to drown the scientific community in a flood of useless information. Reducing the number of scientific journals and limiting their contents to reports of truly significant experiments would not only improve the ability of scientists to keep abreast of significant developments but would also do much to discourage unnecessary experiments that are unlikely to be of any public benefit.

J. *Federal funding of research should be based on the merit of the proposal, not on the reputation of the researcher and the institution.* Not every idea that comes from an elite institution or a well-known researcher merits support. Proposals should be reviewed initially with names of both the individuals and the institutions involved omitted so that the actual research proposal can be considered and rated on its own merit in relation to public health priorities apart from those other factors. The ability of the researcher and the facilities of the institution can then later be taken into account separately. This will permit more objective evaluation of research proposals.

K. *Most research is improperly directed toward obtaining a greater knowledge of scientific issues rather than toward rapid application of research information to solve immediate and pressing public health problems.* This country presently has an enormous wealth of basic health science information that could either immediately or in the near future be put to use to alleviate a great many public health problems. Too often basic scientific research has been conducted as an end in itself, with the result that the practical aspects that could be of great benefit to the public have been obscured or lost.

Federal research institutions currently have little or no interest in applied research as contrasted with basic research. Scientific reputations, academic tenure, and Nobel prizes depend upon the latter, not the former. The federal government must therefore either reorient the basic structure and mandate of its research institutions, or develop new institutions that will fill the current gap and shorten the time between discovery and practical application. It is essential that a major portion of the funds currently directed toward basic research be diverted instead to assuring rapid "technology transfer" to meet the nation's health needs. The public should be spending its funds not for scientific information, but for the ultimate public health benefits that will result from it.

L. *The total lack of coordination on research policy within the federal government results both in duplication of work under federal funds and in promising leads falling through the cracks.* There is no central repository for information on health science research being undertaken or funded by the federal government. Federal research agencies frequently have no information on what other researchers inside or outside the government are doing. A central coordinating function to disseminate information on current research is badly needed.

Rapid dissemination of information about research projects is impeded by the highly competitive and secretive nature of academic health science research. To the extent that federal funds are used in conducting research, any form of secrecy is wholly unjustified and should be forbidden.

M. *Researchers are more interested in the stability and continuity of research funding than in the relevance of their work to the public health and welfare.* Scientists continually complain that, if they cannot depend upon a stable and continuous source of federal funding, their work will be seriously impeded.

Long-term research projects in the federal government, and grants to outside researchers, provide at least a short-term stability that is the envy of many others in society who must be concerned about their economic welfare on a daily basis. Business and individuals can suffer severely by wholly unforeseen circumstances in the marketplace. Privately funded research conducted by industry or through other private funds is therefore constantly at the mercy of both the national economy and specific economic forces that may affect a particular industry, company, or individual.

It is undoubtedly important that health science research not be interrupted or impeded unnecessarily. But researchers have no greater right to economic stability, or to continued use of public funds, than any other citizens. They must therefore learn to cope with the same economic uncertainties and imponderables as the rest of society. The federal government should not hesitate to reduce the amount of funding in health science research proportionately whenever the country must reduce its other economic expenditures.

N. *Special interest groups have too often exaggerated the importance of particular health problems through adroit lobbying, resulting in inappropriate research priorities.* The "disease of the month" approach, although in disrepute, is still not thoroughly discredited. Special appeals are frequently made to the government to give inordinate recognition to a particular health problem without any attempt at an objective determination of its priority among all public health problems. Federal agencies should ignore any special appeal that fails to show its relationship to the overall public burden of illness in this country, based upon objective factual determinations.

O. *Little research is directed toward substantiating the contribution of new drugs and medical devices to improvement of the public health and they are therefore approved without adequate assurance of public benefit.* The Food and Drug Administration is required by law to assure the safety and effectiveness of drugs and medical devices, but not to assess their impact on health care as a whole or on the quality of life. Too little attention has been paid to resulting increases in the cost of medical care and the loss of human dignity accompanied by use of such measures as life support systems that maintain human life for no reasonable purpose.

These are not areas of research that represent the cutting edge of health science knowledge. They are relatively mundane and are unlikely to attract significant scientific analysis without the impetus of targeted federal funding. Federal agencies should therefore divert funds from basic research fields that are well-covered and will attract major investigation to areas of this nature which have an enormous impact on every citizen but are unlikely to receive substantial scientific attention without the inducement of federal funding.

P. *Little research is directed towards substantiating the contribution of medical procedures to improvement of the public health and they are therefore adopted without adequate assurance of public benefit.* Physicians and other health professionals are authorized by law to engage in any form of medical procedure that they believe acceptable. The only present means of controlling unacceptable medical practice—malpractice litigation and revocation of the license to practice medicine—have had little effect in reducing unethical, ineffective, unsafe, and costly medical procedures. It seems doubtful that the medical profession can itself adopt an effective scheme of medical procedure assessment. Accordingly, the federal government should fund research designed to result in a systematic assessment of both current medical procedures and future medical procedures as they come into use.

Q. *Little research is directed toward ascertaining, much less reducing, the cost of health care to the public.* Although Congress and the public are most concerned about the rapid escalation of health care costs, health services research is underfunded, the subject of little interest. Without substantial federal support of this type of applied research—at the expense of basic health science research, if necessary—the country will reach a crisis on health care costs that will dwarf all other health science issues and could result in major reductions in federal spending for basic health science research.

The cost of health care is not presently the concern of any federal research or regulatory agency. Health science research has left this area to economists and policy-makers without adequate thought of its impact on the public. Federal agencies should make certain that in the future cost considerations are taken into account in basic health science research to the same extent that they must be reflected in the policy options available to the health care system as a whole.

R. *Little research is directed to issues of life style and behavior that have by far the greatest impact on public morbidity and mortality.* Even the most obscure disease, which affects only a handful of individuals, receives greater research attention than do the most widespread and important aspects of personal life style and behavior that affect an individual's health. The public health consequences of diet, exercise, sleep, and our other daily habits have long escaped the interest of health science research. Yet the only realistic way of affecting many of the health problems that plague our society is by changing these and other personal habits, and thus the entire way that we live. Federal funding must make certain that health science research focuses strongly on these issues.

S. *Little research is directed toward ascertaining and removing the environmental contaminants and conditions that contribute significantly to human disease.* Some researchers have contended that up to 90 percent of all cancer can be attributed to "environmental" causes. We know extremely little about these causes, and very little health science research today is directed toward them. It is imperative that major research be undertaken to determine the environmental causes of cancer, to reduce those causes wherever possible, and to provide accurate information to the public on the basis of which individuals can make their personal choices that could significantly affect their chances of avoiding cancer.

T. *Little research is directed simply toward improving the quality of life in large populations whose basic condition cannot be changed (e.g., the aged and the infirm).* Regardless how much the fields of science progress in the foreseeable future, one cannot anticipate that the aging process will be halted or reversed. Since this gradual deterioration is one condition which no one can avoid, the federal government should place major emphasis on research designed to improve the quality of life as aging occurs. Much of this research will undoubtedly fall within the area of health services research, but a great deal will also fall within the behavioral sciences and some in the biomedical sciences.

U. *Basic research has made only a minor contribution to the improvement of public health in the past and is not likely to lead to major improvements in the future.* Recent information has shown that morbidity and mortality have substantially decreased because of such general factors as genetic selection, improved nutrition, hygienic engineering, and an overall improvement in basic societal conditions, rather than as a result of increased scientific information attributed to health science research. Further improvements in the future are likely to come from similar societal sources. Federal funding of health science research should therefore be reduced substantially or eliminated, and these funds should in the future be devoted instead to reducing poverty, the main source of health problems in this country.

II. The Public Should Have Greater Control Over, and Protection From, Health Science Research

A. *Health science researchers have no constitutional, legal, or moral right to federal funds.* The use of tax money for health science research is solely within the discretion of the general public. Accordingly, the public may attach any conditions that it wishes to the use of federal funds for health science research, and may change those conditions or reduce or eliminate such finding at any time for any reason it believes to be sufficient. Once a health science researcher agrees to accept federal funds, he also agrees to accept, and to abide by, whatever requirements and conditions are attached to those funds no matter how ill-advised or arbitrary they may seem.

B. *The role of scientists should be limited to conducting the type of research that the public concludes to be relevant to its needs (like the role of generals in conducting wars).* The public is not interested in funding health science research that is merely of academic and intellectual significance. It wants research that is conducted with its money to be of direct and immediate importance in alleviating the nation's serious health problems. Accordingly, the lay public should have the controlling voice in determining what type of health science research should be undertaken. Scientists must then pursue those general areas of interest if they choose to use federal funds to support their research.

This does not mean that scientists in general would be restricted in their freedom of inquiry. Any scientist should be free to conduct, with his own or other private funds, any type of health science research that interests him (as long as it does not pose a danger to the public).

C. *Freedom of scientific inquiry exists only with respect to thought, not with respect to any investigation or action that may affect any person.* Any aspect of scientific inquiry that extends beyond pure thought and writing is therefore fully subject to public regulation and control to prevent potential harm. In this respect, academic research, whether basic or applied, stands on no different footing than commercial research.

D. *Researchers attach too little importance to the protection and the privacy of human subjects.* Examples of unethical human experimentation have been documented in both the scientific literature and congressional hearings. It is therefore apparent that regulatory control mechanisms must be established to protect the public from unethical and improper health science research and to assure that the privacy of those individuals who participate as subjects of such research is securely protected.

E. *The public must be protected from medical procedures which are of unproven safety and effectiveness.* Federal regulation of human experimentation with medical practice and procedures is as important as federal regulation of human experimentation with investigational drugs and medical devices. New laws must be enacted to assure appropriate public control of existing medical procedures whose uses have never been adequately supported and any new medical procedures that may be considered in the future.

F. *Even basic nonclinical research today is involving work that threatens the health of laboratory workers and the general public and therefore requires new governmental controls.* To the extent that basic research in health science may pose dangers to laboratory workers or the public, either directly or indirectly, it should be subjected to appropriate public control through federal regulatory mechanisms.

G. *The present peer review procedure is nothing short of an "old boy" system designed to perpetuate established ideas and institutions and to resist innovations that challenge accepted procedures.* Public participation is essential in the determination of which research proposals have the greatest potential for meeting public health needs. Younger scientists, and those who represent less-established concepts, should also be included in the process. Peer review should become a truly open and public mechanism for determining which proposals better suit national needs, rather than the closed and academic exercise that presently exists.

H. *The present peer review system permits review committee members to make awards to each other's institutions while pretending to avoid conflicts of interest.* It is simply insufficient for peer review committee members to absent themselves from a meeting where a grant to their institution is considered. Steps must be taken that will preclude the type of continuous conflict of interest that arises today. Individuals must be found to serve on peer review committees who have no direct or indirect interest in the outcome of the deliberations of those committees.

I. *The secret meetings of a review committee, where a scientist who has applied for a grant is discussed without his being present and the grant can be denied without the deciding factors being revealed, is undemocratic and denies due process of law.* At the very minimum, any person who is applying for a grant should be permitted to attend any meeting, and to obtain any memoranda and correspondence reviewing that grant application. Criticism both of an individual's ability and of the merits of a research proposal should be made in front of the individual involved, and the individual should be given an opportunity to respond.

The present method of holding secret meetings, without the applicant being present, allows these important matters to be decided on the basis of conjecture, unsubstantiated opinion, and misinformation. Fundamental principles of fairness require that the applicant be permitted to correct any misimpressions or misinformation that may arise.

The failure of the National Institutes of Health to open up review committee meetings to closer public scrutiny makes it look as though the scientific establishment is trying to hide things. Only by allowing the public to know what is taking place in these meetings can the rights of the individual applicants and the broader interests of the public be protected.

III. Federal Funding of Training for Health Science Research Is an Inappropriate Use of Tax Funds

A. *Many graduate students trained with federal funds have available sufficient private funds to support their training.* At present, federal funds for research training are awarded solely on the basis of the ability of the individual, regardless whether that individual has available other funds that could be used to finance that training. The richest and the poorest receive the same amount of money.

Students often have parents who have supported them through college and spouses who can work, or they have access to private sources of funds, such as savings accounts, trusts, insurance policies with cash value. Moreover, there is no rationale for different criteria for financial assistance at the graduate level than at the undergraduate level, where scholarship funds are almost always awarded on the basis of demonstrated financial need.

B. *Many researchers trained with federal funds subsequently earn incomes that make it reasonable to require them to bear the costs of their own training.* Scientists are not notoriously underpaid. Many rise to high positions in government and in industry. Those who perform particularly useful functions for society receive ample financial reward. Those whose contributions to society are small should not receive any greater financial reward than those contributions justify.

Accordingly, all researchers, including those in financial need, should bear the cost of their own training. This can be done through a wide variety of mechanisms, such as ordinary bank loans or by the traditional American way of paying for their education by working nights and weekends. People who do not have the drive to work their way through school one way or another probably do not have drive needed to make a good research scientist anyway.

C. *There is no greater justification for federal training of health science researchers than there is for federal training of lawyers, architects, or clerical personnel.* Recruitment of able personnel in other occupations is as important to national goals as recruitment of able personnel to work in health science research. The structural soundness of buildings and the fairness of our laws are just as important to society as the maintenance of health.

Professionals in occupations other than health science research are routinely required to support their own training at the graduate as well as the undergraduate level, except for cases of financial need. A democracy cannot countenance an inequitable system under which special federal subsidies are given to graduate students in health science research while graduate students in many other equally important professions not only have no possibility of obtaining the same benefits but indeed must pay taxes to support those other graduate students.

D. *Federal support of health science research projects provides indirect federal funding of research training and thus is sufficient to continue to attract qualified researchers without additional funds specifically for research training.* Because the federal government spends billions of dollars each year on all aspects of health science research, large numbers of people have been, and will continue to be, employed in the conduct of that research throughout the country. Because such research grants and contracts indirectly provide training support for the younger research personnel working on those projects, additional federal funding of health science research training is unnecessary and should be discontinued.

E. *Those institutions which have lost federal training funds for health science research in the past few years have found alternative sources of funds.* The federal government is not the sole source of money for research training. State funds, money from private foundations and industry, and, perhaps most important, the tuition paid by students, are alternative sources for financing research training.

A number of educational institutions have lost federal research training funds in the past few years, without significant diminution of their research training efforts. Accordingly, there can be no justification for continuing similar federal funding in other institutions.

F. *There is no correlation between the quality of training given to a research student and the source of funding for that training.* Those research students who work their way through school, as many have done in the past and continue to do, will get the same education as those who receive federal funding. Indeed, it is highly likely that they will come out far better people for the experience, because they are obviously much more highly motivated than those who lead a softer life on a government stipend. Those who are willing to make sacrifices for their education are showing the type of dedication to science that was once the hallmark of health science research and has been undermined by the ready availability of federal funds.

As long as training remains available for anyone who is willing to make the sacrifices to pay for it, those people who are most attracted to health science research will find a way to raise the necessary funds, and those who would just be coming along for a free ride because it costs them nothing will properly be discouraged from doing so. Discouraging hangers-on will have the additional benefit of freeing senior research scientists from unproductive administrative functions and returning them to useful work in research.

G. *There is no correlation between the quantity or quality of research product by an individual and the source of funding for his research training.* Not all of the nation's most illustrious health science researchers were trained with federal funds. No data exists to show that people who have been trained by federal funds, as contrasted with state or private funds, have a greater research output or have advanced the public health to a greater extent.

Training grants and fellowships have uniformly been awarded to the highest quality personnel. One would therefore expect that those high quality personnel—regardless whether they are trained with federal or with state or private funds—might, on the average, ultimately make a greater contribution to the public welfare. All that any such data might show is that, in fact, highly qualified people have been chosen for training in health science research under federal funding. It could not show that, absent those federal funds, the individuals would have been any less productive.

H. *Federal research training grants are undemocratic because they favor a few elite universities and discriminate against the majority of universities which could use federal funds to improve their programs.* Federal funding of research training programs might possibly be justified as "seed money" to develop strong research training programs in particular universities where other sources of funding have not materialized in the past and the program has demonstrably suffered as a result. This would recognize the national importance of research training without continuously subsidizing programs that are more properly financed through the usual form of student tuition.

The present system, however, provides a limited number of training grants to the very universities which presently have the strongest and best programs and which therefore need the funds least. This policy is creating an even wider gap between a few top research training institutions and their competitors.

Even worse, it results in a general lowering of research training standards rather than an improvement. The best research training institutions will continue their high quality even without federal funds. Those federal funds could therefore be put to better use in other institutions to improve the quality of their research training programs. The net impact would be a larger number of high quality research training programs than presently exist.

I. *Significant amounts of research training funds are used to finance the overhead of academic institutions and to purchase new equipment rather than for the research training itself.* Present regulations permit up to 25 percent of a training grant to be used for this purpose. This only makes the richest universities even richer and further exacerbates the disparities between those who are able to obtain such grants and those who cannot.

J. *A significant number of researchers trained under federal funds pursue nonresearch careers.* Whether this is because these people were simply getting a free ride and had no serious intention of engaging in a research career, or because there are too many people trained for the available positions and other careers must therefore be pursued out of economic necessity, this fact justifies immediate discontinuation of all federal funding of health science research training. At a very minimum, any person who ultimately pursues a nonresearch career should be required to pay back the federal government the full amount of the federal funds expended on his behalf (including any funds used by the training institution for administration and overhead).

K. *Federal funding has produced an oversupply of trained researchers, many of whom cannot find appropriate jobs.* This is attributable at least in part to federal funding of health science researchers, which attracts more poeple into the field than can properly be absorbed.

The number of people who should be trained in health science research depends solely upon the number of people who can be absorbed in the marketplace. In large part, this is determined by the federal health research budget. Since that budget is not likely to increase dramatically, and there is already an oversupply of trained health science researchers, any possible justification for further federal funding of health science research training is eliminated.

It is apparent, indeed, that the artificial stimulation caused by federal funding of health science research training has, by overriding the free market mechanism for manpower in this area, created a serious national problem. The current glut of qualified health science researchers is likely to remain for some years to come even if all federal funding of research training were to be eliminated promptly.

Proof of the seriousness of the current problem is readily obtained from data showing that, even with the reduction in federal funding of health science research training at the predoctoral level in the past few years, the number of such students has nonetheless been increasing. Not only does this dictate immediate elimination of all predoctoral health science research training support, but it strongly suggests that other measures must be taken actively to discourage young people from beginning a career in this area.

L. *The free enterprise market mechanism is the best means of assuring that enough (but not too many) health science researchers will be available to meet the country's needs.* Maintaining the large numbers of trained researchers for whom there are no appropriate jobs now or in the near future on a permanent life-support system consisting of an ever-increasing number of postdoctroal fellowships simply is not the solution. It seems far more beneficial to them and to society, in the long run, to let their research ambitions die so that they may go on to pursue productive careers in other endeavors. To do otherwise would be to maintain a large cadre of highly trained and qualified health science researchers in positions of limited productivity and satisfaction, with an ever-increasing backlog of these people being produced every year.

In the future, therefore, federal funding of all research training should be eliminated. There will of course be lags between the time that the market signals a need for greater or lesser quantities of research personnel, but there is no evidence whatever that this lag will be anywhere near as great as the gap that has now resulted between academic supply and market demand as a result of federal funding of research training in this area. In short, government planning here, as in so many other places, has been shown to be utterly inadequate and inappropriate.

Conclusion

Each of these criticisms, taken by itself, may well be insufficient to cause significant concern in the scientific community. Taken as a whole, however, and repeated daily throughout the country, they must be regarded very seriously and given appropriate attention before they become accepted as the basis for future health science research policy.

The NAS Committee on a Study of National Needs for Biomedical and Behavioral Research Personnel, established pursuant to the National Research Service Award Act of 1974, has begun to analyze on a systematic basis the issues raised in part III of this paper, but a great deal more work is needed before its task is completed. No systematic review of the issues raised in parts I and II appears to have been undertaken, although sporadic attempts to deal with particular aspects of those issues have been made by a wide variety of organizations and individuals.

It is unlikely that Congress or the executive branch of the government will take any action in the short run that would result in substantial change in health science research policy in this country. The scientific community therefore has an opportunity in the years ahead to pull together a complete and convincing defense of current policy, to make changes to strengthen it where appropriate, and to assure that freedom of scientific inquiry in this area will continue with undiminished vigor in the future.

HARVEY BROOKS

The Problem of Research Priorities

SINCE THE PEAK of federal science funding was passed in the United States in about 1966, there has been evident a growing gap between research opportunities and financial resources in science. In academic science available support measured in dollars of constant purchasing power per eligible active investigator has declined by more than a factor of two.[1] Such a development was inevitable eventually, and had been predicted by many, but it burst upon the scientific world as a surprise after centuries of nearly uninterrupted growth of science and technology. Of course, during most of this time the scientific community was so small that its claim on economic resources was scarcely noticeable, so that science could advance almost independently of the state of the economy. Even in the Depression of the 1930s the national scientific enterprise continued to grow in both funding and employment, in each year of the decade except 1932–34. The average annual real growth was between 7 percent and 9 percent between 1930 and 1940.[2] Thus, an actual decline in support and manpower is a relatively new phenomenon, even though particular institutions, such as government laboratories, have suffered severe temporary cutbacks.[3]

When science funding was increasing, questions of priority received little overt notice because worthy new projects and new investigators could be supported with little detriment to work already under way. It is only recently that competition among projects has become so severe that the scientific community has become aware of many worthwhile projects which could not find support and many able scientists who could not find professional opportunities commensurate with their training. Some of this, of course, is part of the general phenomenon of expectations rising faster than resources in our society, but some of it represents a real constriction of resources in relation to opportunities. As a result, there is a feeling that the cost to scientific progress arising from wrong choices of scientific direction may be considerably higher than in the past.

Priorities in science cannot be set without bringing in considerations external to science itself, especially if the projects compared are only distantly related. There is nothing inherent in scientific logic which can say that molecular biology is more important than elementary particle physics, or cosmology more important than evolutionary biology, although some distinctions can be made on criteria such as degree of generality, philosophical implications, or elegance and simplicity. These, however, are primarily aesthetic rather than rational criteria. In the past, the relative priority accorded to many disciplines, such as physics or electrical engineering, has resulted from their connection with certain national "political" missions, such as defense or the development of nuclear energy. Although this connection may originally have had an intellectual justification, the evolution of the field has in many cases outgrown this connection. Low energy nuclear reactions, for example, were rather closely related to nuclear reactor and nuclear weapons design; but out of this early work in nuclear physics there evolved the subject of high energy particle physics, whose connection with energy was limited to the use of the word "energy" in labeling the field. The changing priority of national missions has been reflected in the allocation of resources to the basic scientific fields originally attached to them. Thus it is that real support for the physical sciences has dropped by nearly a factor of two since 1967.[4] Public disenchantment with the military-space-atomic energy syndrome of the past quarter century has thus been visited on the fields of pure science which rode on its coattails for so long. When the only public function to which the electorate seemed willing to devote large resources with no questions asked was national security, it was natural to justify the support of science under this label. The magnitude and complexity of the defense budget, and the tendency of Congress to concentrate its attention on the big hardware items, meant that research received little detailed scrutiny, and this became a device for protecting the internal autonomy of science. As the subject which benefited most from association with the military-space emphasis, physics is now the field that has suffered most from public disenchantment.

In a time of reordering of social priorities, it is natural for the public to question priorities even in basic science, often without understanding the looseness of the connection between the intellectual and social priorities. The defense–space umbrella which protected some fields of basic science now serves as the basis for questioning their continued priority.

Another pressure for the establishment of scientific priorities arises from the belief that science is inherently "rational" and that therefore science, of all activities, should be most capable of developing rational criteria for choices among fields. Why shouldn't the setting of research priorities be as "scientific" as science itself?

One part of the problem is, of course, the confusion between "science" and the total federal R & D budget, which are often treated as interchangeable terms in popular and newspaper discussion. Most public discussion of priorities is in terms of large technological projects such as the SST, the space shuttle, or the fast breeder reactor, although such technological projects have only a peripheral relation to the development of the scientific disciplines, despite the fact that they consume more than 80 percent of the funds expended on R & D. The argument has been still further confused in the last two decades by a tendency to pursue some major technological projects largely for their own sake, rather than for a clearly defined social purpose. Atomic energy and space technology were pursued for a variety of social purposes which tended to change with time. The technical content of the programs was much more constant than their stated social purposes. For a while, it appeared that even the ocean sciences would be pursued for their own sake in their own special federal agency.[5] Such efforts might be classified as "pure technology," by analogy with pure science.

They were technologies pursued in an integrated way in terms of their own inner logic for a variety of loosely defined political purposes. The confusion is even further compounded by the large component of purely technical apparatus development associated with some areas of pure science such as high energy physics or space astronomy. How is the layman to distinguish between the importance of a manned space flight and the importance of a giant particle accelerator? Yet, in the eyes of the scientific community the accelerator is only ancillary to the asking of a fundamental question about nature, while in the case of manned space flight, scientific knowledge is a mere by-product of an activity pursued with a largely political purpose.

One of the central issues of the growing debate about scientific priorities is who should be involved in the process, and how? To what degree should the scientific community itself have central responsibility for the determination of the broad strategy of science as well as its more detailed tactics? Increasingly, there has been a debate between the public and its representatives on the one hand, and leaders of the scientific community on the other, as to where the line between strategy and tactics should actually be drawn.

The scientific community perceives that the strategic decisions about science are governed by budgetary and economic considerations and by political appeal, and that the views of working scientists receive a hearing only at the margins. The needs of the disciplines, for example, are seldom weighed against the requirements of "pure technology," as became apparent in the battles between ground based astronomers and the space agency a few years ago. Scientists watch hundreds of millions of dollars committed to technological spectaculars such as Apollo or the SST, while major scientific facilities are underutilized for lack of a few million dollars of operating funds.

On the other hand, political administrators point to hundreds of millions of dollars expended on biomedical research by the National Institutes of Health (NIH), largely under the control of study sections and advisory panels made up of scientists whom they view as beneficiaries of these same expenditures. They observe strategic choices being made on the basis of criteria which, although they allude to the cure of disease, are governed more by the internal logic of scientific inquiry than by the possible use of the results. Many politicians and some "nonestablishment" scientists believe the priority setting processes of science to be dominated by self-serving elites oblivious to democratic accountability and larger human concerns.

This particular set of issues has come to a focus in the recent debate over the priority given to molecular-level investigations in biomedical research and more specifically in the debate between the "environmentalists" and the "molecular biologists" in respect to priorities for cancer research.[6] In Europe, the issue is being joined even more sharply in the agitation over public participation in the formulation of "science policy," as described by Dorothy Nelkin.[7]

The Total Resources of Science and Technology

A national science policy is a policy for the allocation of scientific resources. It can be discussed in various domains ranging from basic science or academic science alone to the whole field of research and development, or indeed in the total process leading from research to fully commercialized or deployed technology. Technological, managerial, and social innovation form part of a seamless web; categorization of any part of the process as a separate activity is partly artificial, and for the same reason, a clear separation between social priorities and scientific priorities can also be only approximate.

A number of observers have pointed to the lag in United States' expenditures for R & D during the last decade in comparison with such expenditures in other advanced industrial countries.[8] In part, this may be closely tied to a lag in capital investment by United States' industry, which in turn reflects both a decline in rates of personal saving and a decline in after-tax profits of industrial firms. Thus, estimates indicate that between 1965 and 1976, profits after tax, with adjustment for inflation and also with fixed capital adjusted for its replacement cost, dropped from 11.8 percent to 4.8 percent.[9] Since the results of most industrial R & D are embodied in investment goods rather than consumer goods, it is scarcely surprising that the decline in investment is reflected in declining industrial R & D. Similarly, in government, emphasis on social programs, which are analogous to consumer expenditures, has been reflected in declining federal R & D investment in real terms.[10]

One of the central issues of science policy is what proportion of the national product should be invested in R & D, and how this should be divided among basic research, applied research, and development. Whether in fact this is a meaningful question is a matter of intense debate. Many industrial research directors, for example, stress the artificiality of the distinction between basic and applied research so far as industry is concerned, and administrators of medical research often take a similar view. In the 1960s, the director of the NIH tried to avoid any classification of biomedical research as basic or applied. Similarly, many students of science policy argue that the aggregate R & D cannot be determined apart from the particular menu of social and political goals that specific R & D programs are intended to support.

There is, however, at least one sense in which total resources going into R & D are meaningful from a policy standpoint. At some level of expenditure, the productivity of the expenditure is limited by the supply of talent available. This situation may have obtained during the period from about 1955 to 1965, when salaries of scientists and engineers rose appreciably relative to wages and salaries generally, there was a large influx of scientific and engineering manpower from abroad, and many technicians and skilled workers in industry were upgraded to engineers despite a lack of formal educational qualifications.[11] But since the midsixties, R & D activity has been limited by financial rather than human resources, as indicated by a steady fall in relative salaries, as well as by the statistics on support per investigator mentioned previously.

A possible science policy would be one which treated scientific and technical manpower as the scarcest resource. Spending on research and development, or possibly on various subcategories of R & D, would then be based on the amount of talent in each category, and the question of priorities would reduce to the question of the allocation of talent. In support of this approach, one might quote the substantial body of economic research that indicates that the average social returns to R & D are considerably higher than returns to investments in physical capital.[12] By this argument, it would be advantageous to invest more money in R & D provided a sufficient level of talent is available and the mix of projects is chosen so as to maximize the total social return within a given level and mix of manpower.

The foregoing line of argument seems to presuppose a rather sharp threshold of talent and motivation for scientific research. In this view there is a pool of such potential talent in each age cohort whose proportion of the cohort is quite inelastic with respect to available research funds, provided quality control is retained through peer review. The contrary view is that there is a continuous grading off of talent and motivation and that, in the long run, as more money is invested in R & D, the talent drawn into it is increasingly marginal, and it is impossible in practice to define the point of diminishing returns. The peer review process itself tends to modify its standards to match the available talent.

A second argument against the manpower allocation view is the simple economic argument that research on the productivity of R & D investments has given evidence only on *average* returns, but what is significant for deciding on increases in investment are *marginal* returns, i.e., the incremental return per incremental dollar invested. This economic argument is especially cogent if indeed there is not a sharp threshold of qualified talent.

These arguments tend to be supported by the observation (particularly in basic research) that research productivity among scientists is highly skewed; most of the significant research publications are produced by a small minority of scientific workers.[13] Derek Price has argued, for example, that aggregate scientific output tends to rise as the square root of the number of active scientists,[14] indicating that marginal productivity goes inversely with the square root of the total number of scientists engaged.

The above criticism might be partially answered by adding a constraint on the average salary of scientists and engineers. The policy would be to keep the total technical work force fully engaged without a rise and fall in the average "price" of technical people relative to other forms of labor. The work of R. B. Freeman suggests that in the private economy the price elasticity of technical employment is quite high[15] in the long term, about 0.7 (that is, a 7 percent decrease in employment for a 10 percent increase in average compensation).

In a manpower allocation policy such as that sketched above, federal R & D expenditures would serve primarily as a "balance wheel" to private R & D. Total federal R & D support would be adjusted gradually to bring the total technical man years of effort to some target value related to the total number of trained people available. The problem of priorities would then consist of the selection among competing projects that could be carried out with the residual pool of manpower not fully absorbed by private industrial R & D.

In contrast, the present de facto policy in the United States is that the scarcest resource is not technical talent but discretionary dollars in the federal budget. The limit on aggregate federal R & D expenditures is set by complex trade-offs among competing demands for the federal dollar, especially social welfare and military spending. When competing demands were relatively low, and available revenues rising rapidly, as in the early sixties, R & D expenditures rose so rapidly as to place inflationary demands on technical resources. By contrast, in the seventies, discretionary federal dollars were eroded by competing demands and failure of federal revenues to keep pace, and this resulted in many critical scientific and technical resources being left underutilized. Because of the cumulative nature of scientific and engineering skills, including the long educational path, as well as their rapid depreciation with disuse, the R & D system cannot adjust efficiently to changes either in priorities or in total expenditures. Hence, the underutilization of technical manpower may have a particularly high social cost.

Indeed, the trade-offs tend to be made on a sector-by-sector basis rather than overall; thus, defense R & D competes with defense expenditures in general; agricultural R & D with the Department of Agriculture expenditures in general, and so on. In fact, the trade-offs are so many-faceted that it is probably misleading to speak of them in terms of any single binary competition.

Both the manpower and fiscal models of aggregate R & D expenditures imply global R & D budgets, the one expressed in terms of technical man years of effort, the other in terms of money. Resources are to be suballocated in such a way as to maximize some sort of output or a weighted combination of outputs constituting a single-dimensional criterion of merit. In the manpower allocation, it is assumed that, because it cannot fully capture the benefits of research, the private sector will tend to underinvest in R & D from the standpoint of maximizing the social benefit potentially derivable, and therefore government supplementation is required.

Most proposals for allocation of a global research budget focus either on academic research as an aggregate or on basic research, assuming that applied research and development activities will reach a natural level determined by competition with other operational social objectives. The argument for singling out academic or basic research is that its success is more dependent on long-term stability of support. However, it is impossible to ignore the impact of a fluctuating external employment market on the conduct of academic or basic research. Thus, there is some requirement for overall stability in the R & D enterprise as a whole.

The manpower allocation model corresponds rather closely with the practice in the Soviet Union, where the output of the system of advanced education is fine-tuned to the number of jobs available in the whole R & D enterprise, which in turn is funded according to a gradually increasing total determined under the current five-year plan for research and development. Thus, the entire Soviet R & D system is based on what can be described as a "level of effort" approach, with the total funding being adjusted to keep the scientific work force fully occupied, something which is possible in the Soviet system because the entire enterprise, including salary scales, is closely controlled centrally by the government. Planning for the agricultural research system in the United States follows a somewhat similar pattern. The problem with such a system is that it tends to lead to rather low mobility and flexibility. The level of support is tied to institutions whose budgets are quite predictable from year to year, and the allocation of resources within the institutions is made centrally in terms of only gradually changing priorities.

The manpower model has several attractive features, but runs a large risk of inflexibility. If such a system of allocation is to retain vitality, a scientist might have greater assurance of productive employment than at present, but should probably have less assurance of employment in exactly the place or activity he preferred. There would have to be greater adaptibility to changing priorities. In the Soviet system, it is very difficult to gain access to the system, which is very competitive in its early stages, but once access has been gained, a scientist who succeeds is likely to stay in the same institute and career line for his entire working life. In the United States, access is easier, but forced exit is much more probable, and the frequency of mobility and career change has been much higher, at least until recently. During the period of tightening funds and the disappearance of university expansion, a more complex pattern may be emerging. Insecurity and mobility may have increased in the early career stages, but the institution of tenure, combined with the rapid expansion of the recent past, may be introducing a rigidity which did not exist earlier. Although the system has a high degree of adaptability, many feel that it produces too much wastefulness at the margins. There is a growing consensus in the United States' scientific community that a better compromise between stability and flexibility is possible and desirable, especially for the first decade or so of a scientific career.

Truth versus Utility

The federal government supports research for essentially pragmatic reasons—because it is an important means by which agency goals can be achieved more effectively or more efficiently. In other words, social utility is the primary criterion for federal support, and for choice among areas of support.[16] The operational consequences of this principle, however, are by no means obvious. For one thing, the scientific and technical enterprise needs a certain measure of internal autonomy in order to pursue even the most pragmatic goals. This is true because the ends we can pursue are constrained by the possibilities presented to us by nature, and in the words of Charles Fried[17] we cannot affirm "truth as a constraint on the pursuit of the useful, while denying truth any power to set its own agenda." In addition, "only the pursuit of truth for its own sake can develop and maintain the standards which are necessary to ensure the objectivity and the value of any research, pure or applied."[18] Even in the most applied research activity, "once the search is underway, utility is bracketed and reality must be the goal, lest the desired utility itself never be reached."[19] These principles, in fact, express the implicit agreement between science and society which was first formulated by Francis Bacon, and which has served well both the growth of knowledge and the realization of its utility.

The frequently demonstrated usefulness of "useless" research is sometimes used as an argument against any kind of social guidance of research. Like the "invisible hand" of the classical free market, it is argued that the autonomous working of the intellectual free market of ideas will produce socially optimal results at the least cost. The most eloquent exponent of this view was Michael Polanyi, who coined the term "republic of science" as a symbol of the idea of scientific autonomy.[20] But this concept scarcely seems viable in a polity in which a large fraction of the resources invested in the enterprise of science comes from the public, and in which the total investment is as large as it is in the modern industrial states.

By now, most politicians concede the necessity for some degree of self-governance and internal agenda setting in the scientific community, and most scientists concede the necessity for some political and public input to the setting of the general directions and goals of scientific research. The issues are where the lines should be drawn and the appropriate processes by which the scientific and political communities should mutually negotiate the scientific agenda. For the most part, it is conceded that scientists cannot determine social goals, and politicians or the public should not determine scientific methods or tactics, nor influence conclusions. The problem is how to reach agreement on what constitutes ends and what constitutes means, where the line between strategy and tactics in science is to be drawn. The problem is frequently formulated in terms of the relative roles of internal and external criteria in setting research priorities.

The institutional expression of science's need for autonomy is the so-called peer review process. In principle, this is a process for using expert judgments to select the best projects within a defined field of science in terms of "scientific merit," that is, in terms of criteria internal to science. The cardinal principle of peer review is that judgments of scientific merit should be made entirely by scientists, and moreover by scientists who are active in the same field of research or one closely allied to that proposed by the investigator. Even in the case of institutionalized research in government and industrial laboratories, the fundamental judgments regarding the selection of scientific personnel are made by peers, especially in the most productive laboratories. In this paper, I intend to assume that peer review does indeed select the "best science," and that there exist more or less universalistic criteria of scientific merit which any competent group of scientists will agree upon, at least in application to concrete scientific proposals. There appears to be ample empirical evidence from recent research by sociologists of science that such "objective" criteria of scientific choice do exist; therefore, the merits of this proposition will not be debated further here.[21]

The legitimate controversy over the merits of peer review applies not to its validity in rank ordering projects or scientists within a field, but to its broader use in the formulation of science policy and the allocation of resources *between* fields of research. Peer review is a process which operates most successfully when it is used only in the "truth" dimension and not in the "utility" dimension of scientific choice. In fact, it is common knowledge among students of science policy that peer review has been much less satisfactory when attempts were made to use it to evaluate proposals in applied research, for example, the RANN (Research Applied to National Needs) program of the National Science Foundation[22] or in some of the more clinically oriented programs of the NIH.[23] Even when the practical goals of a research project are agreed upon, there is usually much less consensus among reviewers concerning the merit of a scientific proposal. Frequently the viability of the practical goals themselves is also a bone of contention among different reviewers.

There is another important difficulty with peer review, and that lies in the definition of a field of science. Generally speaking, the broader the intellectual territory covered, the less consensus there will be on the ranking of a group of proposals within that territory. Also, the consensus is usually greater at the extreme upper and extreme lower ends of the quality spectrum. But in the middle ground, which constitutes the majority of projects and scientists, the disagreement tends to widen as the field to be covered widens. Furthermore, many fields of science can be categorized in a way which implies either a predominance of internal criteria or a mixture of internal and external criteria.

Thus, solid state physics and materials science embrace a large area of overlap in actual subject matter, but projects in solid state physics can be judged almost exclusively in terms of truth criteria, whereas, the term materials science may imply a considerable element of utility in judging merit.

In fact, the categorization of subject areas in science presents many opportunities for controversy with respect to peer review; a field is often given a useful sounding label in order to increase its political attractiveness, whereas peer judgments would have been more unanimous if the field had a "pure" label.

However, the peer review process relates to the setting of priorities among fields of science in an entirely different way. It is frequently used as a signaling mechanism to indicate to program administrators where the most significant opportunities lie within a much broader domain of science. This comes about through "proposal pressure," which represents what the scientific community has pluralistically decided is most worth doing, with some sort of additional weighting by the quality of the ideas and people involved in specific proposals as judged by peer reviews. Since the selection of senior scientists in universities itself involves a much broader comparison among distant fields, the proposal pressures coming from university investigators reflect these broad comparative judgments within the academic community.

Of course, proposal pressure can be used as an indicator of scientific opportunity only with a considerable judgmental factor added, for there are many factors which tend to distort it. For example, the mere announcement of the availability of money for certain kinds of research will attract proposals, and this effect will be much enhanced if there is a general shortage of research funds or if funds are only available for a limited number of research areas, with many gaps. Researchers will tend to go where the money is, but if money is more or less equally available in a large number of fields, then those fields with the most significant scientific opportunities will tend to attract the largest number of high quality proposals.

Proposal pressure is also difficult to interpret when projects differ greatly in size or require coalescence of the activities of large groups of people, as in large national or international programs such as the International Biological Program or the International Decade of Ocean Exploration. Although many of these programs involve individualistic subprojects, they have to be closely interrelated and require the use of common facilities and logistic support, and hence, a considerable measure of central planning. This planning is usually done with the aid of committees of scientists closely involved in the general field of work. In some degree this may be regarded as an extension of proposal pressure in the sense that the planning groups are engaged in the identification of scientific opportunities which are thus signaled to administrators.

Another problem with the use of proposal pressure is its tendency to reflect momentum generated in the system by previous actions. Earlier support of certain fields of research generates personal, institutional and even bureaucratic commitments within the supporting agencies which tend to become self-perpetuating. In other words, prior support itself generates proposal pressure which might then be interpreted as an indication of a need for additional support. Moreover, support results in the training of students who may then become future claimants for support in the same field.

One can imagine large numbers of very high quality proposals in a particular field, each one of which makes very good sense by itself, but which are sufficiently alike in objectives and subject matter so that supporting all of them would have dubious justification in terms of the overall balance within a much broader field of endeavor. If all similar proposals were considered simultaneously by the same set of reviewers, the overconcentration of effort might become apparent, but, in practice, proposals are usually received at different times, often by different agencies, and are reviewed by different referees. Comparison of projects tends to occur more automatically with large projects involving expensive instrumentation. A sophisticated program officer or advisory committee can make allowances for the effects of overconcentration; hence, it is difficult to know in practice how serious the effect is. Systems of peer review, such as those used in the NIH study sections, where a committee reviews a large number of proposals at the same time, are probably more effective in avoiding overconcentration of research effort on fashionable topics than systems which involve only individual mail reviews carried out at different times.

Charges have been made that the project system tends to favor "safe" proposals and to reject "off-beat" but highly original projects which do not fit accepted paradigms in the field. It is also charged that it favors experimental work for which there is already a well-developed theory at the expense of "long shot" experiments not favored by theoreticians. It may also favor projects which promise quick, publishable results at the expense of projects requiring much longer to come to fruition. Again, these are hazards which can be compensated for by sophisticated scientific judgment of program administrators and advisory groups, but there exist few data to indicate the extent of validity of such criticisms in practice.

In theory, proposal pressure can be used to compare widely distant fields, since it does not involve much substantive comparative analysis of the scientific content of projects, but only a general accounting of numbers and dollar amounts of proposals previously evaluated as of high quality. In practice, the sociology of research tends to vary from field to field, so that proposal pressure may be difficult to interpret on a comparative basis. Thus, a given number of proposals in organic chemistry may have quite different significance as to the needs of the field from proposals in, say, molecular biology. Proposal pressure is also subject to manipulation by an entrepreneurial program officer who can simply go out and generate proposals among his constituency in order to prove the neediness of the area of science for which he is responsible. However, abuses such as this are usually fairly evident to higher level management in an agency, especially if the strategic decisions are made by managers with scientific training and experience.

All of the considerations above suggest that proposal pressure is an imperfect indicator of the grass roots priorities of the scientific community, and cannot be used as the major guide to the allocation of resources, especially among widely different fields of science. On the other hand, it does constitute an important signaling system that should not be ignored in the priority setting process. When interpreted by scientific administrators with good judgment and wide knowledge of both the substance and the sociology of the scientific enterprise, proposal pressure can be a better indicator than might be supposed. For example, if money is pumped into a new field, and the quality of the resulting proposals is very low, as often happens, this is a caution light on further increased support that ought to be heeded, no matter how politically appealing the field. Conversely, the first indication of emerging new areas of science often comes as the result of unusually imaginative and original research proposals submitted by scientists with outstanding records of prior accomplishment in adjacent areas of research.

The use of the peer review system as an indicator of research opportunities places emphasis on internal criteria, on truth rather than utility. Its use for priority setting implies an implicit priority for the "best science" regardless of considerations of social benefits or consequences. It must be recognized that any other scale of judgment implies a conscious decision to support less good science at the expense of better science for reasons lying outside science.

The question then arises, what should be the role of scientists in dealing with the utility dimension of priority setting, and indeed what should be the role of the public? Clearly, the special skill which scientists can bring to bear is the capacity to predict the probable or possible consequences of proposed research. Although scientists are the only people with the requisite knowledge to do this, it is still an unaccustomed role, and they do not do it very well. When scientists are asked to defend their basic research in terms of utility, the result is frequently naive and forced, a job of salesmanship rather than thoughtful analysis.

Furthermore, the role is a peculiarly difficult one, involving an inherent conflict of interest, especially in a period of constricted research funding. The incentive to oversell basic research on grounds of its utility becomes greater as the competition for research funding becomes keener. At the same time, the excesses of a few entrepreneurs may ultimately discredit the whole exercise of forecasting the consequences of research. The problem is exacerbated by the attitude of the news media which tend to overemphasize highly speculative or uncertain applications of basic research, or, more recently, equally speculative and uncertain adverse consequences.

Furthermore, as we move from applied towards fundamental research, we soon reach a point where the probable results of research, as well as the possible practical consequences of a variety of possible outcomes, become so uncertain and so diverse that the only honest thing we can do is to fall back on a generalized faith in the average beneficence of new knowledge. The problem is that any one practical consequence is so improbable or speculative that it would be dishonest to use it as a justification for supporting the research, and yet the sum total of possible benefits, each rather unlikely by itself, may add up to justify it. In many cases, the information obtained which may lead towards a better future assessment of benefits or adverse consequences of possible applications may be the principal justification. This is essentially what Charles Fried means when he says that truth must be allowed to set its own agenda.

The difficulty today is that neither scientists nor the public have the former unqualified belief that all knowledge is ultimately to man's benefit. In the last several decades, too many new technologies have emerged whose benefits are ambiguous, to say the least, and critics are quick to point out that, although science may not have "created" these technologies, they could not have been created without the spectacular advances of scientific knowledge in the twentieth century. Perhaps one of the best examples is modern chemistry, which has only recently come to be seen by some critics as the ancestor of many industrial products that pollute our environment. As long as one was confident that any piece of scientific knowledge would have beneficial consequences with, say, a 90 percent probability, one did not have to concern oneself with forecasting the consequences too accurately. But this confidence no longer exists, whether justifiably so or not. Pessimism about the consequences of knowledge is part of the general climate of our times, and although scientists are more optimistic than most, more and more of them share in the general pessimism. Indeed, there seem to be as many scientists ready to search out uncertain and highly speculative adverse consequences as there were scientists a few years ago ready to exaggerate the potential benefits of research. In part, perhaps, this is the result of the fact that an increasing fraction of research is "defensive," i.e., devoted to the identification and avoidance of harm rather than to the creation of products and processes intended to satisfy a potential market. Thus, there is a growing domain of research which might be said to have a vested interest in the exaggeration of disbenefits, just as many scientists have had a vested interest in the exaggeration of potential benefits.

There is a complementarity between our ability to assess scientific merit and our ability to assess utility. For a project whose utility is very difficult to assess, and highly speculative, scientific merit is usually easy to assess, and is amenable to peer review. On the other hand, science which is more "applied" may be easier to assess from the standpoint of utility but harder to assess for scientific quality. We have already seen that the process of peer review has run into serious difficulties and criticism in the applied area. There is a large gray area in between "pure" science at one end, and highly targeted applied research at the other, where the difficulty of assessment arises primarily from the inability to agree on the relative weights to be assigned to truth and utility in evaluation. For example, there is much useful research which is regarded as very pedestrian when weighed on the scale of scientific quality. An example is the testing of chemicals for toxicity, the carrying out of chemical analyses, or some kinds of work in systematic biology. These gray areas are usually battlegrounds for academic scientists and scientists from industry or government, with the latter tending to assign greater weight to utility.

Unfortunately (or perhaps fortunately) the complementarity between usefulness and scientific merit is not complete. Research undertaken primarily for purposes of utility frequently turns up fundamental issues which lead to research projects of high scientific merit by anybody's standards. An example is some of the crystallographic and metallurgical work carried out in connection with transistor development. Conversely, very fundamental work leads to vistas of applied work of obvious and great utility, as in the case of the laser. Particularly in the fields of materials science and medical research the interplay between utility and truth has been extraordinarily fruitful.

But in cases where utility is difficult to assess and highly speculative, it is probably better to ignore it altogether as a basis for priority setting. Conversely, where scientific merit is difficult to assess, but utility is clear and certain, one should probably drop scientific merit as a basis for priority setting. This may be what happens in practice, although it is always difficult to "sell" research of no apparent utility. For example, high energy physics and galactic astronomy have been supported on a rather generous scale for decades primarily on the basis of criteria of scientific merit, although there have been occasional arguments that the instrumentation development brought about as a by-product of research in these fields has turned out to have high utility. Still, one would probably not have supported the research primarily for the sake of this "fall out" in instrumentation.

It is thus probably in the gray area in which both truth and utility play a role that the greatest effort needs to be made in improving our ability to assess consequences, and in raising the level of discourse between the scientific community concerned with truth and the political community concerned with utility. Biomedical research is a good example of this kind of area, and the dialogue has begun to be joined.

Who Decides?

Of course, the public, and its representatives through the political process, are the ultimate arbiters of research priorities. But that simple statement has little operational significance. In hardly any other field of activity is there a bigger discrepancy between constitutional and informally understood responsibilities. In practice, the public must delegate a large part of its constitutional responsibility to the scientific community, but for this to happen, the scientific community must have and deserve the trust of the public. For only the scientific community can really manage science. When the public steps in, through the political process, to manage science in detail, it not only destroys science but does not get the utility it is after. Furthermore, the capacity of the scientific community to subvert attempted political management is much greater than in most other fields of endeavor, but the more such subversion is necessitated by ill-advised political management, the greater the mistrust of scientists. This leads to a dangerous vicious circle. Hence, the formal decision-making responsibility for priorities in science is much less significant than the informal interactions by which the internal processes of the scientific community interface with the political and administrative process. This relationship is not very susceptible to discussion in terms of rights and privileges, or freedom and accountability. There is no alternative to a largely self-governing science which is nevertheless responsive to the basic aspirations and expectations of the polity. The debate hinges on what is basic and thus pertains to the political process, and what is tactical and thus pertains to the autonomous governance of the technical community. If trust disappears, then the public will take back the delegation which it has made to the scientific community during the last three decades, a delegation which was eloquently articulated in *Science, the Endless Frontier*.[24] To some extent, it is apparently already taking back some of this delegation. The question is how long the process will continue and how far it will go or should go in the best interests of both the public and the scientific community.

Clearly, one aspect of the relationship between the scientific and political communities is that it should not place too great strains on the intellectual honesty of scientists. If the cost of being honest about one's ignorance of the potential benefits of one's research is the total loss of support, then a kind of Gresham's Law will set in which rewards those scientists most willing to stretch the bounds of intellectual honesty in explaining the benefits of their research, or in minimizing its possible adverse consequences or uses. This is not a healthy situation, and in the long run can only stimulate the mistrust between scientists and the public. Some biologists are now regretting the haste with which they brought before the public their speculations regarding the possible consequences of recombinant DNA research. It is not that the issues raised were illegitimate or not properly matters of public concern, but rather that airing of the debate produced the reaction that the scientists had not gone far enough in advocating stronger restrictions. Similarly, many scientists and engineers in the field of nuclear energy feel bruised by public criticism which they think, justly or unjustly, was the consequence of their greater openness in discussing the details of nuclear technology in public than was characteristic of other older, more accepted technologies. Again the public concerns initially raised by a minority of scientists were not without foundation, but they could not have been so well argued or so well publicized without the existence of a much larger body of technical literature (about nuclear safety and the biological effects of radiation) than existed in almost any other field.

On the other side of the coin, the ever-mounting competition for research funds may lead to an extravagance of scientific promises which can only increase mistrust of science and decrease funding still further.

There is a sort of centralized model of priority setting in science in which it is argued that the political process should decide the social priorities of the country, and then the scientific enterprise should be deployed to match these priorities. To a small degree, this is what happens in the multiagency system for the support of science which we have in the United States. That is, the political process allocates money to health, defense, space, agriculture, environmental protection, and so forth, and then each executive branch agency with these social responsibilities is required to come in with a plan for the R & D which will support these social goals. However, means and ends are not so easily separable in the real world. The public does not know what it wants until it knows what it can get and how much it will cost. Social priorities are constantly being changed by the creation of new knowledge, while at the same time the search for knowledge is being partially redirected by social priorities. Within the last decade, for example, geophysicists began to believe that as a result of basic advances in their science, the prediction of earthquakes should be a possibility within the next decade. As a result, the federal government has created a new research program in earthquake prediction. This does not mean that earthquakes are regarded as any more serious a social problem than they were before. The public's priorities for earthquake research rose sharply only when advances in basic science revealed a new opportunity to do something useful which was not conceivable previously.[25]

This example illustrates why research priorities can only be arrived at as the result of a constant dialogue between the political process and the scientific community. Science not only helps in the fulfillment of social goals, but also generates new social goals, and the latter may be a much more important function than the former, though much less predictable. During the last decade or so, there has been a large acceleration of the dialogue between the public and scientists. In the early 1960s, when science in the political process first began to come to public attention through the creation of the President's Science Advisory Committee, the scientists' role was actually quite narrow and technical, mostly confined to national security policy. It was concerned with the assessment of alternative means to largely agreed upon ends, and even in the field of the nuclear test ban the context was largely the same as that for weapons systems. It is only in the last decade that expertise has entered much more pervasively into the political process, in the Congress as well as the executive. There are many scientists now on congressional staffs; the Congress has a science advisory committee in the form of the Office of Technology Assessment; the General Accounting Office has moved massively into the arena of technology and science policy assessment. The federal government and the Congress are swimming in technical studies on energy, environment, health research, nuclear proliferation, and hundreds of other subjects with a high scientific content. It is all chaotic and halting, not following any discernible plan, and it may seem wasteful and costly to the detached observer. But it is the first phase of an important process, which ultimately will promote a stronger connection between science and overall priority setting in the government, and not just the setting of priorities for R & D in isolation.

The process described in the preceding paragraph is primarily governmental, and still attracts little public attention. The other phase of priority setting relates to the direct role of the public in guiding research priorities—the much discussed subject of public participation. This topic has received much more extensive discussion in Europe than in the United States, though at a more theoretical level.[26]

It has come to the fore chiefly in connection with energy policy, where public concern about the long-term safety of nuclear energy has sparked a wide public debate about alternative energy sources and a shift in R & D priorities towards renewable energy sources and conservation. In this country, there is little question but that the very rapid rise in the ERDA budget for solar energy development and demonstration has been largely the result of popular pressures rather than a scientific consensus. On the other hand, it is easy to overestimate the thrust of public participation by itself. In most instances, such changes in priority as have resulted from public input have depended on political activism from an energetic minority in the scientific community. This is certainly true in the solar energy example.

Indeed, the principal effect arising from new mechanisms of public participation, as provided for example in the National Environmental Policy Act, has been to provide a new forum in which minority views in the scientific community itself could get a wider hearing, often bringing pressure on the scientific establishment to reconsider its own opinions and priorities. Much the same applies in the case of the public debate on recombinant DNA research. Without the participation of scientists, it is unlikely that this would have ever gotten onto the political agenda. Hence, in practice, one may say that public participation has provided an avenue for including a wider range of scientific views in the political agenda. This is probably a healthy development, but it would be misleading to attribute it primarily to public participation per se. It is rather a by-product of the mechanisms created for public participation.

There is a danger here in the sense that insofar as public participation is simply an umbrella for quasipolitical debates within the technical community, the hidden political agendas of the technical actors on one or both sides of the debate may be overlooked because of the general respect in which scientific "objectivity" is held by the public at large. In the short term, this may lend a credibility to some scientific inputs that they do not deserve on their merits. In the long term, it may increase distrust of scientists as political actors, and lead to further invasion of lay judgments into the tactical, as opposed to the strategic, domain of scientific priority setting.

An illustration of this problem may lie in the field of research on aging. For a variety of reasons there has been a considerable upthrust of interest in this field in the last few years and, largely under public pressure rather than scientific influence, it has begun to be accorded a higher priority in health research. Yet, in all probability, this new thrust would not have appeared without a strong input from a small minority of the scientific community. The possibility of the prolongation of life has a great deal of political allure, and cautions put forward by other scientists concerning the possible social consequences of these practical results tend to go unheeded. Few in the scientific community would quarrel with the desirability of fundamental investigations of the physiological process of aging, such as investigations into why normal cells in the body cease to divide after a certain number of divisions. There could indeed be close relations between the phenomenon of aging and the cellular phenomena connected with cancer. But some scientists fear popular pressures may seize upon fundamental advances to push practical applications prematurely without careful assessment of the social consequences. In this case, one will have to question whether the alliance between a dedicated minority in the scientific community and understandable populist pressures will be in the long-range interest either of science or of the public.[27] Some will say it is better to bring this all out into the open and let the public decide, but others will question whether bringing in the public at this stage will not create undesirable risks.

Social Priorities versus Research Priorities

Much public unease about the present allocation of resources to federal R & D in the United States stems from an implicit dissatisfaction with political priorities. For example, 63 percent of federally supported R & D goes to space and defense activities, and even when privately supported R & D is considered, there is a heavy concentration in defense-related work. If development of nuclear power is added to defense and space, 71 percent of federal R & D is accounted for.[28] This fraction has declined from about 92 percent in 1963. Nevertheless, in 1978 the downward trend in defense-space-nuclear research was reversed; the largest *percentage* increase in federally supported *basic* research between fiscal year 1977 and fiscal year 1978 was in the Department of Defense. It is often pointed out that 25 percent of the world's technical manpower is engaged in defense-related research. Nearly one-fifth of all PhD physicists in the United States derive some research support from the Department of Defense.[29]

The question suggested by these facts is whether the present allocation is to be regarded as a distortion of R & D priorities, or whether it is to be considered as merely reflecting political priorities determined through the democratic process. Once we have decided as a nation to devote more than 7 percent of our GNP to space-defense-nuclear activities as a whole, the concentration of R & D in these areas follows almost automatically. All represent highly research-intensive activities compared with most other economic activity. Defense production, for example, requires almost ten times as much R & D per dollar of output than does civilian construction. Shifting technical manpower out of defense-related work would imply either heavy displacement of skilled manpower to nonresearch activities (or technical unemployment) or finding ways to make other civilian activities much more research-intensive.

Some critics of defense spending have argued that the size of the military budget is in part driven by R & D spending in the sense that research makes existing weapons systems obsolete and creates the need for replacement. At the same time, constant innovation, especially by the United States, in defense equipment helps to fuel the arms race. From this, it is argued that arms spending around the world can be gotten under control only by getting military R & D under control. By this argument, R & D priorities *do* drive political priorities primarily through the creation of opportunities that would not otherwise be selected by political leaders. I am doubtful that research is the main driving force for military spending; the argument is difficult to verify.[30]

Within the last two or three years, a very interesting debate has arisen concerning present priorities *within* the field of biomedical research. This debate has several dimensions. The one I want to consider as an illustration of the problem of setting priorities has to do with the relative roles of curative medicine and general environmental factors in improving the nation's health status. Critics of biomedical research priorities are quick to point out that there has been very little improvement in the health status of the American population in the last two decades despite the increasing resources devoted to biomedical research and despite the rising proportion of GNP channeled into health care delivery. The explanation usually given for this is that curative medicine has reached the point of diminishing returns, and that environmental factors, including life styles, diet, environmental pollution, social conditions, and even the quality of working life and the sense of self-worth in work, are much more influential factors in the determination of health than our ability to cure specific disease conditions.[31] Furthermore, most research on curative medicine relates to diseases of middle and old age, where the cure of specific diseases has only a small statistical effect on life expectancy. This critique of the biomedical enterprise comes to a particular focus in the case of cancer, where there is growing evidence of the importance of diet and of environmental contamination as a causative factor. Moreover, since the incidence of cancer increases rapidly with age, cure has little impact on statistical life expectancy, although the ability to cure diagnosed cancers has been improving at a steady rate of 0.5 percent per year over the last thirty years.[32] The implication of these observations for the research enterprise might be a rather massive switch of research effort away from the treatment of cancer and towards the study of environmental factors. More broadly, it implies a shift away from emphasis on the health care delivery system as such and towards environmental and social research having little obvious connection with medicine.

The criticism I have outlined does not question the efficacy of biomedical research when measured in its own terms. There has been dramatic progress in the understanding of specific diseases, and in the development of specific cures as in the case of polio, measles, tuberculosis, and many diseases of genetic origin. But this progress, it is argued, has been only marginally relevant to health. In addition, it has led to the creation of a high technology medicine which absorbs an ever growing proportion of health care resources without concomitant benefits in the reduction of mortality or morbidity. There has been little or no relationship between the effort in either research or care technology investment and the potential contribution to improved health. We have sought to push the frontiers indiscriminately in every direction that opportunity presented without reference to cost or to potential health benefits measured on a population basis. We could afford this opportunistic, or "technology push," approach when health care absorbed a minor proportion of national income and when choices of health technologies were made primarily in the marketplace, but in an era when health care costs are rising much faster than the cost of living index, and when a growing proportion of costs are being paid for through insurance or tax money, we can no longer afford to ignore priorities either in health care investment or in R & D.

The same issue seems to arise in health care as in defense. Is the expense and general direction of investment in the health care system driven by R & D priorities, or are R & D priorities driven by the nature and evolution of the health care delivery system? In other words, should the criticism summarized in the preceding paragraphs be directed at the research enterprise itself, or should it be directed at our overall social and political priorities in the health area? If our social priorities were changed, would not R & D priorities adjust more or less automatically to them? Does the momentum of the present system arise from R & D or from the health care establishment itself? Undoubtedly, the answer to these questions is not an either/or one. The coupling between R & D and the social delivery system to which it relates operates in both directions. The question is therefore whether changing R & D priorities would by itself have sufficient leverage on the system as a whole to effect the desired change in our social priorities. I am dubious that it would. Furthermore, one could argue that the current debate over social priorities in the health system is itself largely the product of research findings. A process of change may be taking place in health priorities, driven by research findings, but rate of change must be commensurate with the growth of our knowledge.

It must also be pointed out that as we look at *fundamental* research in the biomedical domain it is not so clear that the critique calls for a radical change in priorities. Probably the most important advance in recent years in our ability to detect carcinogens in the environment has come from the development of bacterial tests for mutagenesis, typified by the Ames test.[33] This is a development made possible by recent advances in fundamental biochemistry and cellular biology. Even for the future, it is not apparent that the shift of interest from curative to environmental factors in disease calls for radical shifts in the priorities for research at the fundamental level, although some such shift may be called for.

What the two examples discussed above seem to show is that the debate over R & D priorities cannot be conducted about R & D as an isolated and separate activity. Furthermore, I do not believe that R & D by itself has sufficient leverage on social priorities to make it a particularly fruitful area for discussion, especially if one is talking about priorities among research areas at the very fundamental level. Clearly, fundamental research should not be guided solely by internal criteria and changes in emphasis with respect to the technologies of health care, and more broadly health status improvement will and should affect relative emphases even at the fundamental level. Nevertheless, it would be a great mistake to think that we can have much influence on social and political priorities by conscious tampering with fundamental research priorities in accordance with external criteria. The most important activity is to maintain vigorous communication and feedback all along the chain between basic research and ultimate application to society's problems, but not to operate the research enterprise by "command and control" from the center.

References

[1]B. L. Smith and J. J. Karlesky, *The State of Academic Science: The Universities and the Research Effort* (New York: Change Magazine Press 1977), chap. II, pp. 9–48.

[2]V. Bush et al., *Science, the Endless Frontier*, reprinted July, 1960, National Science Foundation (Washington, D.C.: U.S. Government Printing Office, 1960) table 1, p. 86.

[3]Charles Weiner, "Physics in the Great Depression," *Physics Today* (October 1970): 31–37.

[4]Smith and Karlesky, *op. cit.*

[5]H. Brooks, *The Government of Science*, (Cambridge, Mass.: MIT Press, 1968), pp. 31–32.

[6]Cairns, "The Cancer Problem," *Scientific American*, 233, (November, 1975): 64–78; cf. also Maguire, A., "Toxic Substances: A View From the Hill," p. 43, in Technical Information Project, *Toxic Substances and Trade Secrecy*.

[7]D. Nelkin, *Technological Decisions in Democracy: European Experiments in Public Participation* (Beverly Hills, Calif.: Sage Publications, 1977).

[8]National Science Board, *Science Indicators 1974*, National Science Foundation, (Washington, D.C.: U.S. Government Printing Office, 1975).

[9]T. E. Mullaney, "Optimism in the Midwest," in column, "The Economic Scene," *New York Times* (Sunday, December 11, 1977).

[10]W. H. Shapley, D. I. Phillips, and H. Roback, *Research and Development in the Federal Budget FY1978*, (Washington, D.C.: American Association for the Advancement of Science, 1977), p. 139.

[11]J. H. Hollomon, and A. E. Harger, "America's Technological Dilemma," *Technology Review* (July-August, 1971).

[12]E. Mansfield, "Contribution of R&D to Economic Growth in the United States," *Science*, 178 (February 4, 1972); Mansfield, E., "The Effects of R&D Expenditures on the Economy," Testimony before Committee on Science and Technology, U. S. House of Representatives (April 28, 1976).

[13]J. R. Cole, and S. Cole, "The Ortega Hypothesis," *Science*, 178 (October 27, 1972): 368–374.

[14]D. J. de Solla Price, "The Science of Scientists," *Medical Opinion Review*, 10 (1966): 88–97.

[15]R. B. Freeman, *The Market for College Trained Manpower: A Study in Economics of Career Choice* (Cambridge, Mass.: Harvard University Press, 1971), chap. 7; Freeman, R. B. and Brenneman, D. W., "Forecasting the Ph.D. Labor Market: Pitfalls for Policy," National Board on Graduate Education, National Academy of Sciences (April, 1974).

[16]Shapley et al., *op. cit.*, pp. 17–18.

[17]C. Fried, "The University as Church and Party," *Bulletin of the American Academy of Arts and Sciences*, 31 (December, 1977): 38; Barber, B., *Science and the Social Order* (New York: Collier Books, 1962), p. 139.

[18]*Ibid*, p. 41.

[19]*Ibid*, p. 38; cf. also Brooks, H., "Applied Research: Definitions, Concepts, Themes," in *Applied Science and Technological Progress*, a report to the Committee on Science and Astronautics, U. S. House of Representatives by the National Academy of Sciences (June, 1967), especially pp. 22–26.

[20]E. Shils, "A Great Citizen of the Republic of Science," *Minerva*, XIV, (Spring 1976): 1–5; cf. "Can Science be Planned?" chap. 3 in Brooks, H., *The Government of Science*, (Cambridge, Mass.: MIT Press, 1968).

[21]S. Cole, L. Rubin, and J. R. Cole, "Peer Review and the Support of Science," *Scientific American*, 237 (October, 1977): 34–41.

[22]H. Simon, et al., "Social and Behavioral Science Programs in the National Science Foundation," National Academy of Sciences, Washington, D.C., (1976).

[23]G. M. Carter, "Peer Review, Citation, and Biomedical Research Policy: NIH Grants to Medical School Faculty," RAND Corporation R-1583-HEW, (December, 1974).

[24]V. Bush, et al., *op. cit.*, pp. 31–33.

[25]Shapley et al., *op. cit.*, pp. 26–27.

[26]D. Nelkin, *op. cit.*

[27]B. L. Neugartin, and R. J. Havighurst, (eds.) "Extending the Human Life Span: Social Policy and Social Ethics," NSF/RA 770123, National Science Foundation, Washington, D.C. (1977), pp. 23 and 26.

[28]Shapley et al., *op. cit.*, pp. 23 and 26.

[29]D. Bromley, et al., *Physics in Perspective*, vol. I, (Washington, D.C.: National Academy of Sciences, 1972) p. 848.

[30]H. Brooks, "The Military Innovation System and the Qualitative Arms Race," *Daedalus* (Summer, 1975): 75–98.

[31]J. H. Knowles, "The Responsibility of the Individual," *Daedalus* (Winter 1977): 57–80.

[32]D. L. Levin, S. S. Devesa, J. D. Godwin, and D. T. Silverman, "Cancer Rates and Risks," U.S. Department of Health, Education, and Welfare, NIH 75-691, (1974).

[33]J. McCann, E. Choi, E. Yamasaki, and B. Ames, "Detection of Carcinogens in the Salmonella/Microsome Test: Assay of 300 Chemicals," *Proceedings of the National Academy of Sciences*, 72 (1975): 5,135.

DOROTHY NELKIN

Threats and Promises: Negotiating the Control of Research[1]

IN AUGUST, 1976, a citizen's review board was appointed in Cambridge, Massachusetts, to consider whether recombinant DNA research would have an adverse effect on public health. The board, composed of nonscientists, described itself as a "citizen's court." Its task was to review the safety procedures set out by guidelines from the National Institutes of Health (NIH) and the methods for monitoring their compliance. Evaluating conflicting arguments offered by scientists for and against the facility, the board presented its findings to the City Council of Cambridge as a basis for local policies governing the conditions of research.[2]

During the same month, a presidential task force proposed an experimental science court to help resolve the factual dimensions of scientific disputes. Like the Cambridge group, the Science Court would arrange debates among opposing scientists, and evaluate their arguments on controversial technical issues. But, unlike the citizen's review board, it would be scientists, not laymen, who would make the judgments to serve as a basis for policy.[3]

The assumptions behind these two efforts were similar; both the Cambridge Experimentation Review Board and the Science Court were intended to provide a rational basis for policy choices in troublesome areas of science and technology; both involved elaborate procedures to accumulate a range of conflicting viewpoints on questions of risk. Yet each represents a different approach to the resolution of disputes. Proponents of the Science Court assume that a major source of controversy lies in disagreement about the nature of evidence. They suggest that neutral judgments by scientists who can distinguish facts from values in controversial areas will help to resolve disputes. The Cambridge Review Board, assuming an inextricable relationship between factual disagreement and value conflicts, argued that "decisions regarding the appropriate course between the risks and benefits of potentially dangerous scientific inquiry must not be adjudicated within the inner circles of the scientific establishment."[4] The contrast between these approaches highlights a crucial question: who should ultimately be responsible for decisions about controversial scientific and technical issues? In the recombinant DNA controversy, this question bears directly on the issue of freedom of inquiry as increasing demands for public control conflict with expectations of scientific autonomy.

The scientific community has persistently resisted public control. Only when research has direct technological applications are scientists willing to concede the need for regulation. In a survey of 800 scientists, seventy-seven percent agreed that "The pursuit of science is best organized when as much freedom as possible is granted to all scientists." This study cited some characteristic attitudes: "A pure scientist must not deny himself a discovery by worrying about social consequences." "I would insist that no area of investigation be closed because someone feels that society is incapable of handling it."[5] Wide consensus on the importance of autonomy has meant significant federal patronage of science with minimal public control. Even as political controversy has constrained technology, control over research has largely remained within the scientific community. More and more, however, concerns about technology are spreading to science. Arguing that basic research as well as its application may have undesirable environmental or social consequences, critics call for greater public scrutiny and sometimes control over research. This prospect is especially problematic for a community organized as a self-governing social system.

What has led to these demands for public involvement in decisions about research? What forms can such public participation take? And what are the implications of these demands for science? The recombinant DNA dispute casts some light on these questions as its participants grope for solutions to a difficult set of problems. But we must begin by briefly framing the case in the context of the present relationship between science and the political community with respect to the notion of freedom of inquiry.

The idea of constraining research calls forth accusations of "McCarthyism," "Lysenkoism," "the Scopes Trial," "the Inquisition." Many scientists feel that freedom of scientific inquiry is a constitutional "right"—like freedom of speech. "The accepted position is and should be jealously to guard the constitutionally guaranteed freedoms, both those expressed and those implied."[6] The concept of freedom of inquiry has a venerable history and is widely taken for granted, but there is no constitutionally guaranteed right to pursue knowledge or to engage in scientific inquiry.[7] Rather, it is more accurate to view freedom of scientific inquiry as the product of continued negotiations. An implicit compact emerged after World War II from the debates over postwar research policy.[8] These debates first focused on the continued role of the military in the management of research as scientists bargained for less restrictive civilian control. Then, Senator Harley Kilgore, concerned with stabilizing the wartime productivity of science, proposed establishing a federal superagency to coordinate scientific work and to finance academic research. Many scientists objected, fearing that federal subsidy would threaten scientific freedom and prolong the wartime pattern of centrally directed research. Especially problematic was Kilgore's notion that the agency should be run by laymen whose appointments could hinge on political criteria. Backed by professional societies, Vannevar Bush negotiated for an agency run by scientists themselves. The key and controversial point in this negotiation was the extent of political control to be associated with federal patronage of basic research. Like Kilgore, scientific leaders wanted stable funding, but only with safeguards against political interference. The negotiation thus hinged on questions of professional rather than lay control and on the degree of direct accountability to the political system and to the president. Truman favored Kilgore's more politically responsive scheme, and it was not until 1950 that Congress finally approved the Bush proposal, insulating the National Science Foundation from direct political control and thereby establishing the basis for a policy that provided federal subsidy for science with minimal public intervention.[9] Science, claims one observer, is "the only institution for which tax funds are appropriated almost on faith, and under concordats which protect the autonomy . . . of the laboratory."[10]

This postwar contract was indeed unusual in its provisions for scientific autonomy and it has been subject to continued erosion precisely at the original points of contention, that is, over the role of laymen in governing the direction of research, the scope for political control, and the extent of accountability to the political system. Nixon's change of mind with respect to appointment of Franklin Long to the National Science Foundation directorship; the proposed Bauman Amendment to the NSF Appropriations Act, seeking congressional control over research projects; the Proxmire "Golden Fleece" awards; the increasing public controls over research methods that involve human subjects—all suggest the fragile character of the relationship.[11] As a symbol of the growing tension between science and the political community, the recombinant DNA controversy has further eroded the contract and brought pressures to renegotiate the relationship and especially the terms of scientific autonomy.

A negotiated agreement has been thoughtfully defined as an exchange of conditional promises and threats.[12] The scientific community has, in essence, bargained for substantial autonomy by claiming the inherent efficiency and, indeed, necessity of an unregulated scientific enterprise, and by promising practical contributions to economic progress in return for funding without intervention. Its bargaining strength comes from belief in the intrinsic value of knowledge, and from the promise of its contribution to the public good. Underlying the negotiation is the implicit threat that society will lose out on the benefits of science if excessive intervention accompanies government support. If the acquisition of basic knowledge is restrained by externally imposed limits, it is society which will bear the costs.[13]

Negotiated agreements are maintained only if both parties have something to exchange that is valued. As long as the products of research were thought to be unequivocally desirable, the premises of negotiation remained stable. However, in the context of growing concern about the risks of research, attitudes have changed, bringing pressures for greater public control.

These pressures reflect a broader ambivalence about the value of science and technology and the legitimacy of decision-making authority. Technical decisions in many areas are confronting public scepticism, expressed in legislative or administrative controls, in legal action, and sometimes in overt protest. Former assumptions about "professional accountability" are increasingly translated into demands for direct "participation." It will be useful for our discussion of the negotiation over recombinant DNA to distinguish these two concepts, for the notion of accountability traditionally does not imply participation. Professional groups, for example, are answerable to the public and expected to account for their actions, but they regulate their own activities. Self-regulation, however, depends on public trust; it is the declining confidence in the adequacy of accountability procedures that inevitably leads to the concern about participation.

It is also useful to distinguish among various concepts of participation. The prevailing assumption is that citizens participate by choosing those they wish to represent them in government. Direct participation in policy-making is limited, and representatives are accountable mainly through periodic elections. Alternatively, it is argued that direct participation by affected interests is necessary in a democratic society. Awareness of the negative effects of certain policies, and a general declining trust in authority, have triggered interest in more active and direct modes of participation. During the negotiations over recombinant DNA research, we shall see both concepts of participation exercised as elected representatives grope for ways to reinforce their legitimacy and competence to deal with a new set of issues, and citizen groups seek more direct involvement.

A final concept in need of clarification is that of regulation. Regulation occurs typically either through economic incentives (funding to encourage useful activities or penalties to discourage harmful practices), or through direct intervention (setting standards or prohibitions). Scientists, like all groups, seek to control the regulatory process, arguing that this is in the public interest. Their ability to do so again rests on trust. Indeed, trust is the key variable in the evolution of the recombinant DNA dispute as declining confidence in the adequacy of conventional self-regulatory procedures in science has led to greater restrictive regulation and demands for more direct public involvement.

I will examine these changing patterns in the negotiation over the autonomy of science. Further, I will suggest the problems of accommodating the trend towards greater public control, drawing from aspects of the recombinant DNA dispute.[14]

Negotiating Premises

Negotiating relationships depend on the attitudes and expectations that are brought to the bargaining process. That a laboratory research technique has provoked such broad political interest reflects changing attitudes about both the scientific enterprise and the appropriate locus of decision-making authority.

CHANGING IMAGES OF SCIENCE

Hecklers at the National Academy of Sciences forum on recombinant DNA opened the meeting singing "We shall not be cloned . . ." They waved flags reading "Don't tread on my genes" and "We will create a perfect race." They questioned the desirability of seeking knowledge with possibly harmful applications and the morality of employing research procedures that pose potential risks. Techniques of isolating and recombining fragments of DNA molecules have opened the possibility of greatly expanding the understanding of genetic inheritance. This knowledge could generate significant advances in the treatment of disease, but it also removes some of the obstacles to genetic engineering. It is this possibility, with its value laden implications, that accounts for the very emotional reaction to recombinant DNA research.

Realistic or not, genetic manipulation is an overriding concern for the most outspoken critics of this research. Although many of them are scientists themselves, they are absorbed in the religious and moral implications of a technique that they perceive as allowing the creation and control of new life forms: "Once the genie is out of the bottle, who can control its application?" "Scientists hold our genetic future in their hands." In the long tradition of intellectual criticism of science, these critics are convinced that science is distorting human values. They use anxiety-provoking images of Frankenstein and Faust. They warn of terrorism and of social corruption through the use of genetic information to justify present inequalities: "Is biology a social weapon?" "Can future generations cope with the possibilities science opens up today?"[15]

The recombinant DNA debate has brought to a focus questions that are raised whenever connections are drawn between genetically mediated characteristics and human behavior. Research on the relationship between race and IQ, or between the incidence of an XYY genotype and criminal behavior, or on the genetic basis of alcoholism or addiction are bound to be controversial. Do we really want to know the genetic basis of intelligence or of behavior? Would this not allow the development of pernicious mechanisms of social control? If one identified XYY individuals or related IQ with race, would not this result in labeling and social stigmatization only to create the very behavior anticipated? Moreover, do we want the ability to control human qualities—to specify physical or mental characteristics? Anxiety about possible applications of research findings leads to questions about research itself. Ironically, it is the success of science, the expectation that knowledge is a compelling basis for social policy that encourages such concerns.

The long-term concerns about the applications of research brought emotional intensity to the recombinant DNA controversy. But the dispute was actually triggered by another issue; the fear of risks inherent in the techniques of research. This fear is not confined to the biomedical context; one might recall, for example, the community opposition to reactor research at Columbia University in the late 1950s.

In the case of recombinant DNA, it was feared that an experiment could inadvertently create genetic changes in known pathogens, or produce novel and dangerous forms of infectious microorganisms for which people have no resistance and medical science no cure. This possibility, if true, could be catastrophic. There was serious concern that risks were enhanced by the prevailing use of *E. coli*—a common bacterium that resides in the human digestive system—as a host for recombined genes, although the common practice was to use enfeebled strains of *E. coli* that lack the capacity to colonize the human gut. The widespread research application of the technique because of the relatively low cost was expected to increase risks. Such fears have been exacerbated by the invisible nature of the risk. How does one know if a lethal gene is produced? It could take several years and the wide introduction of a dangerous gene in the population before problems were detected. Even as increased evidence has shown that accidents are unlikely, fears persist.

One biologist has speculated that the technique may also lead to violations of "genetic barriers" with profound and irreversible impact on natural evolutionary boundaries.[16] Critics point out that the record of containment of biohazards is not flawless even under the most rigorous procedures.[17] They emphasize the magnitude of uncertainty, the "total ignorance" about novel, self-perpetuating organisms.

Some prominent scientists, including James Watson, argue that these concerns are ill-founded. However, in a society deluged with warnings about risks from cyclamates, PCB's, freon, and nuclear power, such fears are inevitable.

"We must control all these experimentations that are taking place and the effects that are coming out of sardine cans and tuna fish cans and even the milk you drink . . . I don't like to be contaminated," asserted one public official.[18]

Realistic or not, as the image of science changes, the case for unfettered research loses negotiating strength. For critics are increasingly asking questions about research that follow more from its external impacts than its internal dynamics. Does one place priority on what can be learned from a research project or on its possible negative social consequences? If risks and benefits are both hypothetical, which is to receive greater emphasis? But most crucial to the negotiation, if there is a conflict between scientific interests and public concerns, does one rely on professionals to assess and control risks in their own research or on those who may be affected?

CHANGING IMAGES OF AUTHORITY

Assumptions about the relative importance of technical competence as a criterion for legitimate decision-making authority are changing, as diverse groups claim social or political competence to evaluate the significance and acceptability of risk, and to decide on policies long regarded as technical and in the province of scientific expertise. The actors in the recombinant DNA dispute include several interests, each with different claims to legitimate decision-making competence based on their role as experts, as citizens, as affected interests, or as officials responsible for the public welfare.

Many laboratory researchers argue that scientific expertise is necessary and sufficient to assess problems of safety. Thus they seek to proceed with research under the old rules, relying on the judgment of their peers and resenting any external regulation as an intrusion on their autonomy. However, a group of molecular biologists at a Gordon Conference in 1973 expressed concern about potential biohazards in the technique of recombinant DNA research, and they proposed that the National Academy of Sciences and the National Institutes of Medicine evaluate the risks and take appropriate action. While these scientists acknowledged, and indeed called public attention to, the issue of biohazards, they too argued that expert competence is the basis of authority and that risks should be contained through procedures established and monitored by the scientific community. Thus, they supported the existing mechanisms of control that rely on self-regulation, seeking only procedural modifications.

A quite different perspective has been brought to the dispute by scientists representing such groups as Science for the People, and the Coalition for Responsible Genetic Research. Many of these scientists were politicized during the Vietnam War when they organized to oppose military research in universities. After the war the focus of their activity shifted to nuclear power and has now turned to genetic research. They start from a very different ideological base. Viewing the problem of recombinant DNA in political and moral terms, they feel that expertise is not a sufficient basis for authority. They question the ability of the scientific community to evaluate the risks of their own research and seek, not simply procedural adjustments, but basic systemic changes in the traditional organization of science and, indeed, of society.

Other actors in the recombinant DNA controversy include the many spokesmen for special interests, ranging from the pharmaceutical industry to labor organizations concerned with occupational safety for technicians. As "affected interests" these groups also claim legitimate decision-making authority. Finally, public officials from federal bureaucracies, regulatory agencies, Congress, and state and local governments have perceived the issue as part of their responsibility to protect public health and safety.

Confronting such challenges, biologists have naturally sought to maintain as much autonomy as possible. But this has only reinforced their identification as a political interest group with powerful career concerns. Indeed a persistent theme throughout the dispute has been the conflict of interest implied by scientists regulating their own research. Metaphors have proliferated. To expect scientists to evaluate the risks in their own research is to ask "incendiaries to form their own fire brigade," or "General Motors to regulate automobile safety," or to ask, "Would the tobacco industry limit the manufacture of cigarettes?" These questions explicitly challenge the assumption that scientists can transcend their narrow private interests and career concerns in making decisions which affect the public welfare. Scientists, it is argued, lack the moral authority and legitimacy to regulate themselves when their work could shape the future of society. Direct societal intervention is necessary to assure that the public interest is properly served.[19]

The Negotiation Process

The changing images of science and appropriate decision-making authority have shaped the negotiation over recombinant DNA research, driving the issue to an increasingly public arena. In response to the concerns about biohazards expressed by biologists in 1973, the National Academy of Sciences formed an investigating committee which, in an unprecedented decision, called for a voluntary moratorium on any research that would improve the antibiotic resistance of bacteria and on any recombinations using tumorous or animal virus DNA. The committee recommended that the NIH develop a program to evaluate hazards, create guidelines for research, and organize an international meeting on the subject.[20] That meeting was convened at the Asilomar Conference Center in California in February, 1975.

The Asilomar Conference was organized as a technical discussion of the scientific issues that defined biohazards and experimental safety.[21] It was intended originally as a means of educating the international scientific community about the hazards of working with potentially pathogenic organisms. The organizers had invited a few members of the press, but were soon overwhelmed by the public's extraordinary interest in their meeting, for many scientists did not regard the issue as a matter for public scrutiny at all. It was for scientists themselves to judge the extent of acceptable risk, and to take proper safeguards in the public interest. The important point for most scientists in this professional context was to manage the problem of risks so as to proceed vigorously with research that offered the promise of substantial benefits.

The meeting was motivated by concern that "If the collected wisdom of this group doesn't result in recommendations, the recommendations may come from other groups less well qualified."[22] Thus, the participants agreed on interim guidelines based on the proposition that physical and biological containment must match the hazards involved, and sent recommendations to the NIH with the expectation that their efforts would be acclaimed as a model of social responsibility. "You have seen scientists at work trying to do what they felt was right for the public, not for themselves," one scientist later testified.[23]

Nevertheless, Asilomar did not mark the end of controversy. The first to react was a group of scientists who had been concerned for some years with "the social responsibility of science."[24] These activists questioned whether, in the climate of intense competition, one could trust scientists to comply voluntarily to recommendations that would constrain their work. Would it not be in the best interest of individuals to avoid compliance? Scientists might hesitate to call attention to problems that could bring external scrutiny. From their critical perspective, Asilomar appeared less a model of social responsibility than a means to protect vested interests.

The translation of the Asilomar recommendation into NIH guidelines met some but not all of this criticism. NIH has budgetary ties to Congress and responsibility to elected officials. Through its administration of research funds in the area of molecular biology, the agency represents an authoritative institution able to set standards for research to which grant recipients are accountable. Policy decisions in the NIH are set by a council which includes public representatives. However, NIH decisions about research rely on the peer review system, so that its role in setting guidelines exacerbated concerns about self-regulation and further polarized the scientific community. The first draft of the NIH guidelines, weaker than the Asilomar recommendations, was promptly attacked by critics as "a means to protect geneticists not the public."[25] Deluged with letters, the NIH added stricter requirements, only to be criticized by traditionalists as "terribly responsive to outside lobbying rather than being objective and thinking about it."[26] However, the new guidelines, describing requirements for physical containment during experiments and prohibiting any research endowing bacteria with toxins or antibiotic resistance, were released.[27]

By laying out containment requirements, the NIH established standards of protection, and criteria for accountability. The procedures used to develop the guidelines, however, presented little challenge to the pattern of self-regulation and thus intensified concern about decision-making authority. Although the NIH meetings were open to observers and the committee received public testimony, the guidelines were written by scientists involved in research and concerned primarily with providing for safety without limiting the range of inquiry.[28] Some observers expressed doubts about the discretionary nature of the procedures: "It is extremely unlikely that Congress and the public would be willing to rely solely on the moral suasion engendered by the guidelines and the peer pressure that they could carry . . ."[29] Others questioned how the NIH, with responsibility to promote research, could also regulate it. They argued that the measures of accountability were insufficient, for the primary responsibility to assess potential biohazards still rested with principal investigators or local biohazards committees composed mainly of professionals likely to concentrate on facilitating research.

If dissenters cannot win their battles at one level, they typically seek to broaden their constituency.[30] Thus, opposing scientists sought a wider framework for negotiation through greater public involvement. The negotiation quickly moved beyond the scientific community to engage environmentalists, public interest groups, congressmen, and local and state government officials.[31] The Environmental Defense Fund and the Natural Resources Defense Council petitioned the Department of Health, Education and Welfare (HEW) to hold hearings, arguing that the affected public had not had an adequate opportunity to participate in decisions about research with broad social and ethical implications. Friends of the Earth called the guidelines "a violation of the democratic process," and called for more public representation in the formulation of research policy in the hopes of avoiding the "tragedy of the nuclear industry."[32]

In response to such pressures, several congressional committees held public hearings. Three major questions dominated congressional discussion: how to cover industrial research under NIH guidelines, whether or not to establish a national regulatory commission, and whether federal legislation should preempt state or local controls. There was wide agreement on the need for federal legislation to extend controls over industry, but the institutional questions have been controversial. Senator Edward Kennedy proposed a new national commission with powers that would overlap those of the NIH by licensing facilities, inspecting laboratories, and monitoring compliance to regulations. Others opposed the proliferation of new federal agencies. Kennedy's bill would allow state and local government rulings to prevail despite federal legislation. Other proposed legislation based on the NIH guidelines would preempt local community rulings.

When local government officials became aware that research was planned in their communities, they too entered the dispute. In 1976, the Harvard University Committee on Research Policy approved plans to upgrade an old biology laboratory to meet the new NIH standards. However, alerted by a newspaper article and by local activists, the mayor of Cambridge, Alfred Vellucci, a long time antagonist of the academic establishment, brought the issue to the city council. In two tense days of public hearings the council heard testimony for and against Harvard's laboratory.[33] Discussion focused on enforcement procedures. Those satisfied with the NIH guidelines emphasized the technical adequacy of the containment procedures; others claimed, "The guidelines in and of themselves are meaningless unless we have some police force, hopefully of a non-biased point of view, which has the interest not only of the lab technician but the general public in mind." With this in view, the government of a city which would bear the costs of research in case of accident, asserted its political authority to enter negotiations about research.

Mayor Vellucci's hearings were rather more colorful than the usual discourse about science, yet his perspective reflected the prevailing view of the role of local government: that public servants have the responsibility to intervene in matters affecting the health of a community. Vellucci reminded scientists of his obligation as a representative "to make sure nothing is being done in the private or public laboratories that may be injurious to the health of the people of the city."[34] The city council had forbidden the construction of slaughterhouses as dangerous to the public health and the mayor suggested that the city should have the same say about a research laboratory: "Why should one separate one health hazard as against another health hazard?" Obliged to enter decisions affecting his constituency, he felt that many scientists deliberately obscured their activities with unnecessarily technical language. "Refrain from using the alphabet," he told scientists at the public hearing. "Most of us in this room including myself are lay people; we don't understand your alphabet." He complained about secrecy of decisions about research: "If it wasn't for some of these newspapers we wouldn't have known nothing about this stuff—we caught Harvard just in time."[35]

Finally, Vellucci reminded scientists that public officials have a right to intervene in activities supported by public funds. Reflecting a more widespread concern with accountability (think of Proxmire's monthly "Golden Fleece" awards and the proposed Bauman Amendment) Vellucci asked "Who the hell do scientists think they are that they can take federal tax dollars and do research work that we cannot come in and question?" He offered a political definition of academic freedom. "They don't pay taxes and so they are free taxable institutions and that's all the freedom they are gonna get . . ."[36]

The city council established a moratorium on research in Cambridge, and asked that a review board evaluate the adequacy of NIH procedures. The city manager selected people from a cross-section of the community; all were nonscientists with no connection to Harvard or MIT. They included a former mayor, a nurse, a community activist, a Tufts University professor of urban policy, a former city councillor, a physician, and a social worker. The review board was more a jury than a representative body; indeed, board members saw themselves as a citizen's court. They met four to six hours a week for four months, and members claimed to put twice that time into "homework," educating themselves on the technical issues by reading articles (an estimated twenty pounds of reprints) visiting laboratories, and hearing seventy-five hours of expert testimony. In a mock trial, they examined the views of opposing scientists. Although most participants began with the assumption that any suspicion of risk should preclude research, in the end all agreed that the research should continue. They proposed, however, additional monitoring procedures, including broader public representation in the university biohazards committees required by the NIH and a Cambridge biohazards committee that would oversee research in the city.[37] In February, 1977, the city council supported the board's recommendations, which then became an ordinance.[38]

Coming to Terms

POLARIZATION AND MISTRUST

At least three levels of dialogue are woven into the dispute over recombinant DNA research: a technical discussion of safety, a philosophical discussion of values, and a political discussion of authority and trust. For most biologists, the issue is one of safety to be controlled by careful containment procedures; for critics, recombinant DNA is a sensational technology with ethical and social implications requiring greater public control. For many lay observers, this debate is a symbol of declining trust in science and its governing institutions to represent public values—"People don't trust the authorities to run this thing; we have lost our capacity to trust."[39]

As the dispute evolved, drawing increased public attention, attitudes polarized nearly to a point of noncommunication. The very use of language reflected the diverse perspectives of those engaged in the dispute. The scientists at Asilomar, calling for greater self-regulation, had initially talked of "recombinant DNA technology" and even "plasmid engineering." Later, when threatened with external control, their language shifted, and they referred to the research as "a science" directed to basic understanding of the human genotype. Critics, on the other hand, consistently called it a "technology." There were other linguistic manipulations. Several biologists suggested that the *E. coli* used for recombinant DNA research be given a new name so as not to confuse it with unattenuated strains. A scientist critical of the research objected to the use of the word "molecule" for DNA fragments since it implied chemical rather than biological manipulation.

Threats and promises escalated. Especially bitter about the schisms within the scientific community, biologists repeatedly emphasized the benefits of their work and warned that excessive public control would retard research at enormous public cost. "It will be contrary to the public interest if this should lead to a decision by the public to direct the scientific course of such investigations."[40] The same scientists who opened the discussion about recombinant DNA at Asilomar had second thoughts as the issues became politicized. They perceive their critics as irrational and hysterical, committed to the destruction of science, or at best possessed of a Chicken Little mentality that amplified speculative risks into expectations of disaster.[41] However, when biologists pointed to the limited possibility of an accident, their critics argued that the possible magnitude of an accident, no matter how unlikely, was an overriding consideration. When biologists claimed the benefits of genetic research for future medical technologies, critics responded that this claim was exaggerated because disease is related more to environmental than genetic factors. Most scientists felt that regulation, appropriate with respect to technology, should not be extended to basic research where future applications are unpredictable. However, their critics argued that concerns about technology inevitably bring into question the implications of knowlege itself, and that the moral implications of genetic research as well as immediate risks call for public intervention. Finally, the most profound disagreement centered on the extent to which scientists themselves can be trusted to evaluate the consequences of their research, and impose their own constraints.

THE PARTICIPATORY IMPULSE

For those who are unwilling to trust the authority of governing institutions or established administrative procedures, direct participation becomes a means to assure accountability. The NIH guidelines were not in themselves a sufficient guarantee that scientists would be accountable for the hazards of their research, so critics demanded procedures for more direct public involvement and local control. The mayor of Cambridge clearly took special pleasure in comparing Harvard to a slaughterhouse; but his perspective about the need for greater public involvement in science policy is more widely shared. The mayor of Ann Arbor, Michigan, argued, for example, "I will not abdicate my responsibility as mayor of the city to any federal, state, or private institution within our political jurisdiction. Just as scientists must reexamine the whole question of academic freedom and right of inquiry, so must the various units of government reexamine their respective goals and relationships." The mayor went on to call for a greater public voice in monitoring and evaluating research with ethical or social implications.[42]

Similarly, an attorney, Harold Green, has argued that participation is needed "to stir up controversy," for scientists represent only a narrow spectrum of social values. Science policy, like tax policy or federal fiscal policy, he contends, should be subject to a democratic process including bruising political debate: "I do not see anything that is inherent in science that ought to distinguish it from any other aspect of our society in terms of the operation of the political process. Everything else is subject to the adversary process and debate, why not biomedicine?"[43]

To many scientists, such claims for broader participation and political debate are a challenge to science itself—a manifestation of growing antiscience sentiment. In fact, the Cambridge Review Board decision belies that assumption. Unlike many citizen actions, it did not bring a stalemate to the negotiation. The question of abandoning the research was not considered as board members explicitly decided to focus on the adequacy of containment procedures and the provisions for their implementation, rather than the ethical implications or potential future applications of the research. Although several members in fact were personally concerned about these broader ethical issues, they agreed to follow the instructions of the city manager and to limit their considerations to questions of safety. This decision suggests that these citizens shared the premises of the scientific community about the value of research. Rather than an antiscience demonstration, the participatory impulse in this case represented a search for ways to establish a more appropriate relationship between science and those affected by it. Essentially, the Cambridge Review Board was an experiment on how a lay citizen group could participate in decisions concerning science. The group wrote,

> Knowledge whether for its own sake or for its potential benefits to humankind, cannot serve as a justification for introducing risks to the public unless an informed citizenry is willing to accept those risks . . . we wish to express our sincere belief that a predominantly lay citizen group can face a technical scientific matter of general and deep public concern, educate itself appropriately to the task, and reach a fair decision.[44]

This participatory experiment reflects the changing political context of decision-making in other sectors. The recombinant DNA conflict is only one of the many controversies over science and technology. Similar disputes rage over the siting of nuclear power plants and the implementation of biomedical innovations, as well as over the methods of research. Many decisions once defined as technical have become the focus of adversary politics. In the 1960s, for example, a decision to build a power plant was a matter of closed negotiation between a utility, the industry, and its regulatory agencies. By 1970 changing values required greater public accountability in order to limit social and environmental impacts.[45] Participatory procedures have been integrated into administrative processes; they have been institutionalized in sunshine laws, in provisions for public access to documents, and in strategies to involve the public in decision-making. In lieu of trust in responsible authorities to act in the public interest, these procedures are an adaptive response to demands for accountability in an increasing number of policy areas.[46]

As demands for accountability descend on the scientific community, we find a similar groping for appropriate accommodation. Both the proposed Science Court and the Cambridge Experimentation Review Board can be viewed as efforts to win public acceptance of controversial decisions. Each has its own problems. The task of the Science Court would be to seek "neutral" scientific judgments about the factual dimensions of disputes, assuming that this neutrality would contribute to policy choices. However, this and similar proposals (for computer mediation,[47] or a new profession of certified public scientists[48]) are simply elaborations of the self-regulating mechanisms of science. Supporters of such proposals assume, as did the Asilomar organizers, that conflict is an aberration, failing to recognize that basic value differences underlie disputes over the acceptability of risk. And they ignore the central concern with vested interests when scientists judge themselves.[49]

At the other extreme, the Cambridge Review Board and similar politically based groups are often composed entirely of laymen from the local communities where research is proposed. These groups may have mixed motives, for their deliberations may reflect past experiences (e.g., town–gown relationships) more than the problem at hand. With no scientific representation, they are unlikely to win the confidence of the scientific community unless, as in the Cambridge case, they make a decision compatible with scientific interests.

Other social inventions such as the human experimentation review boards required in institutions supported by HEW funds include both scientific and lay representation. A survey of these institutional review boards found substantive differences between the perspective of the laymen and the scientists who were participating.[50] The scientists emphasized the medical contributions of research, and defined their purpose as weighing the protection of human subjects against the need to develop new knowledge (implying that greater risks can be tolerated in high priority research). The first concern of lay participants was the adequacy of informed consent, and they perceived the purpose of the review procedure to be the protection of the subjects. Although lay members on the boards were generally less active than scientists, and the boards were relatively uncoercive, university investigators by no means fully accepted this mechanism of accountability, in large part because of the lay involvement. Nearly half of those interviewed felt that board members were making judgments beyond their technical qualifications and that the procedures were a bureaucratic intrusion on their work.

THE SCIENTISTS' RESPONSE TO PUBLIC PARTICIPATION

The search for forms of public participation proceeds with frustration and ambivalence, reflecting the persistent tension between the ideal of participation and its pragmatic implementation. As a concept, participation is a source of legitimacy, but as a procedure it may be inefficient and obstructive. Among scientists it is feared that widespread public involvement in decisions concerning science would virtually paralyze the conduct of research. For beyond the usual procedural complications involved in establishing participatory reforms, science poses special problems that are apparent in the recombinant DNA dispute.

To assess research procedures requires evaluating complex technical material. Lay groups have significant problems coping with the knowledge required to evaluate research and may in the end fall back on whatever expertise is available. In large part, the Cambridge Review Board relied on the existing NIH guidelines and on the judgment of one of its members who had both pedagogical and community organizing skills.

Beyond the problems of complexity, however, research that probes the frontiers of knowledge may present risks that are vague or even hypothetical. There is seldom full and conclusive evidence that could serve as a definitive basis for predicting the effect of research, or its potential harmful applications. And even if it were possible to predict certain risks, there is no calculus by which to evaluate their public acceptability.[51] Inadequate understanding of the nature of risk could bring about unrealistic demands for risk-free research. Mayor Vellucci, for example, wanted "an absolute 100% certain guarantee that there is no possible risk which might arise from this experimentation."[52] It was only after months of intense study of the details of recombinant DNA research that members of the Cambridge Review Board, inclined initially to demand risk-free research, realized that this was not realistic.

Greater public participation may also disrupt patterns of behavior accepted as normal for the practice of science. The pressures of responding to the media, the tactical political manipulations, and the local or specific interests of most lay groups are alien to the way scientists perceive themselves. Mayor Vellucci insisted that scientists testifying in the Cambridge public hearings identify their political base: "I represent the lay people of the city, who are you here to represent?" Similarly, hecklers at the National Academy of Sciences forum on recombinant DNA demanded that speakers identify their source of support as evidence of potential bias. In each case the demand created discomfort and irritation among scientists who perceive themselves more a part of an international community than a local political interest group.

Although openness is important to the scientific ethic, the demand for greater public information threatens the normal practice of science. When the City Council of Cambridge wanted access to research proposals, scientists sought to identify them as proprietary and to keep them from open circulation. Suggestions that the Freedom of Information Act be extended to provide open access to proposals horrified scientists who felt that such public access would allow theft of ideas by other scientists in fields where intense competition for priority in discovery leads to self-protection, would disrupt the peer review system and the unbiased evaluation of scientific work, and would provide opportunities for outside intervention.[53]

Aware of these problems, many scientists have responded to the recombinant DNA dispute with dismay or bitter disdain. Fears are "science fiction" in the imagination of "a few ideologues" concerned with furthering their own political views. Critics are "kooks, shits and incompetents."[54] Their views reflect "antiintellectualism" and "political misbehavior." One scientist compared the fear of physical contagion from pathogenic microbes to the seventeenth century fear of moral contagion by soul corrupting books—both reasonable in their way but both pernicious.[55] Many scientists felt public hearings to be unnecessary, and regulation to be destructive—"the first step towards government control," the "camel's nose under the tent."[56]

The overwhelming concern with maintaining self-regulation became explicit whenever regulation appeared to be imminent. In the spring of 1977, for example, the United States Environmental Protection Agency (EPA) sought opinions from professional societies about its potential role in monitoring the research. The Genetics Society of America wrote to forty eminent molecular geneticists, and seventeen replied.[57] Thirteen asserted that research would pose minimal environmental hazards or none and that therefore there was no reason for regulation. Interference would "border on the ridiculous." The EPA should stick to "known hazards such as radiation exposure from nuclear power plants," they believed. The other four geneticists suggested that it was not really known whether or not there would be environmental risks, but two argued that, in any case, EPA should not be involved until the scientific community itself established evidence of risk: "The whole question of regulation and monitoring is abhorrent especially when done by a government agency and not by scientists." Above all, these scientists were concerned with the trend towards local government regulation. Indeed, as the participatory impulse was expressed by the interest of local communities in the issue, uniform federal legislation appeared increasingly desirable.

PRINCIPLES OF ACCOMMODATION

While many scientists remain firm in their resistance to outside regulation, the degree of autonomy from political interference that was negotiated during the period of post-World War II optimism no longer seems feasible in a far more sceptical age. Demands for public involvement and societal control imply new relationships—a new "concordat" which reestablishes the position and status of scientists with respect to the larger society. If the autonomy of science is viewed as a negotiated contract, not an absolute right, several principles follow.[58]

First, to establish viable relationships, bargaining parties must have a sense of political efficacy. Those who feel excluded from decisions they regard as controversial and who have no power to exercise sanctions are more likely to obstruct than to accommodate. In highly technical areas, political efficacy rests on availability of information. The Cambridge Review Board had good access to technical information and conscientiously educated itself. As an informed group with a sense of potential influence, it was able to take a constructive mediating role.

Second, negotiating relationships must be based on some mutual understanding and shared values—an esprit communauté which emphasizes mutual objectives, if only the need to maintain future relationships. The Cambridge Review Board shared with scientists a respect for the scientific endeavor; it focused more on the appropriate means to implement containment procedures than on the value of the research itself. Much greater difficulties of negotiation arise when basic ethical and value differences prevail. The prospect of genetic engineering, like abortion or fetal research, tends to polarize opinion on the basis of conflicting values. Thus, the split within the scientific community, based on concerns about the moral implications of genetic research, appears non-negotiable; for when values are not shared, no facts can be brought forth beyond a certain point to change people's minds.

Third, a viable process of negotiation requires that controversial issues be stated in terms of problems to be solved rather than solutions to be accepted. This implies candid disclosure of the potential risks as well as benefits of scientific research. If research in a sensitive area is initiated as a noncontroversial technical exercise, public mistrust is likely to increase. The aftermath of Asilomar suggests the dangers of disclosing potential risks, but the scientists' initiative may have deflected harsher criticism; if the same questions had been raised by the public rather than by scientists, subsequent constraints on research could have been far more obstructive.

Fourth, negotiations must recognize and deal directly with issues of public concern. This controversy and others reflect more than simply fear of immediate risks. Authority is a major issue as the debate increasingly centers on the choice of appropriate controls. Asilomar demonstrated that scientists are attentive to risks, but it failed to deal with the concern about self-regulation; evidence of social responsibility simply did not answer the lingering questions about the locus of control.

Finally, negotiations must leave open some possibilities for compromise. Compromise is an ambiguous word. It may carry positive implications of agreement by mutual consent, or unsavory overtones of weakness and unprincipled concession. Scientists, unaccustomed to bargaining and compromise, are inclined to take the negative view. Yet there is room for compromise since much of the debate has to do with the level of containment required for specific types of research.

The recombinant DNA controversy has called attention to the importance of establishing a viable relationship between science and society that will allow continuing negotiation about specific problems of research. To simply argue the "right" of free inquiry is unrealistic, for there are already many restrictions on science. Since the 1946 Atomic Energy Act, all research activities using fissionable materials have had to comply to the security regulations and licensing procedures of the Atomic Energy Commission. During the 1960s, federal regulations virtually prohibited scientists from using psychogenic agents to elucidate basic biochemical mechanisms of brain action. Restrictions on the use of human subjects for biomedical research are by now routine. The recombinant DNA issue differs mainly in the extent to which it has become a focus of wide public scrutiny. Given the policy importance of scientific research and its concern with basic life processes, such public scrutiny is inevitable. The negotiation is no longer over whether there will be greater public control of science, but over who will participate in establishing controls, how controls will be organized, and how much they will influence detailed decisions concerning the nature and procedures of research.

REFERENCES

[1]Research for this paper was financed in part by a joint NSF-NEH EVIST grant. I appreciate extensive comments provided by the MIT discussion group organized by G. Holton and R. S.

Morison. In addition, Rae Goodell, Clifford Grobstein, Philip Handler, Sheldon Krimsky, Richard O'Brien, and Vivien Shelanski provided important criticism.

[2]James Sullivan, City Manager, letter to the City Council of Cambridge, August 6, 1976; Cambridge Experimentation Review Board, Guidelines for the Use of Recombinant DNA Molecule Technology in the City of Cambridge, submitted to the Commissioner of Health and Hospitals (December 21, 1976).

[3]Task Force of the Presidential Advisory Group on Anticipated Advances in Science and Technology, "The Science Court Experiment," *Science*, 193 (August 20, 1976): 653.

[4]Cambridge Experimentation Review Board, *op. cit.*

[5]Marlan Blissett, *Politics in Science* (Boston: Little Brown, 1972), chap. 3.

[6]See editorial in *Science*, 192 (September 19, 1975), by DeWitt Stetten, Jr. Also see National Science Board, *Science at the Bicentennial*, (Washington, D.C.: U.S. Government Printing Office, 1976), pp. 59–69.

[7]Harold Green, "The Boundaries of Scientific Freedom," Harvard University *Newsletter on Science, Technology and Human Values*, 20 (June, 1977): 17–21.

[8]The story of these debates appears in Alice K. Smith, *A Peril and a Hope* (Chicago: University of Chicago Press, 1965). See also Daniel Kevles, "The National Science Foundation and the Debate over Postwar Research Policy," *ISIS*, 68 (1977): 5–26.

[9]See Vannevar Bush, *Science: The Endless Frontier* (Washington, D.C.: U.S. Government Printing Office, 1975).

[10]Don Price, "The Scientific Establishment," *Science*, 134 (August 18, 1961): 1099.

[11]For a discussion of those growing pressures, see Dorothy Nelkin, "Changing Images of Science," Harvard University *Newsletter on Public Conceptions of Science*, 14 (January, 1976): 21–31.

[12]F. C. Iklé, *How Nations Negotiate* (New York: Praeger, 1964), p. 7.

[13]Scientists assume that external contraints would have a basic impact on scientific progress. Michael Polanyi argued that "any authority which would undertake to direct the work of the scientists centrally, would bring the progress of science to a standstill," in Michael Polanyi, "The Republic of Science," *Minerva*, 1 (Autumn 1962): 68. Similar arguments pervaded the recombinant DNA dispute. "We could not have penicillin or any of the other benefits of medical research over the past century," stated A. Pappenheimer in "Hearings on Recombinant DNA Research," (City of Cambridge, July 7, 1976), p. 204. See also Stanley Cohen, "Recombinant DNA: Fact or Fiction," *Science*, 195 (February 18, 1977): 654–659.

[14]A comprehensive review appears in Science Policy Research Division, Congressional Research Service, *Genetic Engineering, Human Genetics, and Cell Biology*, Report for Subcommittee on Science, Research and Technology, U.S. House of Representatives (December, 1976). For documents, see U.S. Government, Department of Health, Education, and Welfare, National Institutes of Health, *Recombinant DNA Research* (Washington, D.C.: U.S. Government Printing Office, August, 1976).

[15]The above concerns were vividly expressed at the National Academy of Sciences Forum on Recombinant DNA (March 7–9, 1977).

[16]Robert Sinsheimer, "Genetic Engineering: the Modification of Man," in *Impact of Science and Society*, 20 (April, 1970): 279–283.

[17]Department of Health, Education, and Welfare, *op. cit.*, p. 233. In testimony during the NIH hearings Dr. Emmett Barkley summarized studies of laboratory-acquired infections; there were 3,921 reported cases in the last three decades. At Fort Detrick 423 laboratory infections and three deaths were reported from 1943 and 1970. See National Research Council, "Report of the Panel on Risks and Benefits of Recombinant DNA Research" (November 1, 1977).

[18]Mayor A. Vellucci of Cambridge, Massachusetts, "Interview by NOVA," mimeographed paper (September 4, 1976), p. 19.

[19]Harold Green, "Law and Genetic Control: Public Policy Questions," in Marc Lappé and Robert Morison, *Ethical and Scientific Issues Posed by Human Uses of Molecular Genetics*, Annals of the New York Academy of Sciences, 265 (1976): 173.

[20]Maxine Singer and Dieter Soll to Philip Handler and John Hogness, "Guidelines for DNA Hybrid Molecules," Letter to the Editor, *Science*, 181 (September 21, 1973).

[21]Proceedings of the Asilomar Conference appear in Department of Health, Education, and Welfare, *op. cit.*

[22]Stanley Cohen, quoted in Nicholas Wade, "Genetics: Conference Sets Strict Controls to Replace Moratorium, *Science*, 187 (March 4, 1975): 935. Some scientists, however, argued that making a public issue out of self-regulation would only bring about outside control in any case.

[23]David Baltimore in testimony on February 10, 1976, Department of Health, Education, and Welfare, *Recombinant DNA Research*, *op. cit.*, p. 254.

[24]They were mainly affiliated with Science for the People, a group of scientists who organized during the Vietnam War over the issue of military research in universities. Expression of their concerns appears in testimony by Jonathan King, Cambridge Hearings, *op. cit.* (June 23, 1975), pp. 135–145; Richard Lewontin, *ibid.* (July 7, 1976), p. 188; and Dr. Silverstone, Department of Health, Education, and Welfare, *op. cit.*, p. 286. Proposals have included eliminating the possibility

of a Nobel Prize in this area in order to reduce competition, and restricting research to only a few isolated laboratories.

[25]Jonathan King quoted in Nicholas Wade, "Recombinant DNA: NIH Group Stirs Storm by Drafting Laxer Rules," *Science*, 190 (November 21, 1975): 769.

[26]W. Bennett and J. Gurin, "Science that Frightens Scientists," *Atlantic* (February, 1977).

[27]The guidelines (Cambridge Hearings, *op. cit.*) described four levels of physical containment for experiments, ranging from adherence to standard laboratory practices to requirements for the use of special buildings and extreme decontamination procedures. They also outlined levels of biological containment based on the survival rates of the recombined bacteria use in experiments; and bacteria were to be developed with mutations that would prevent survival outside the laboratory. See Department of Health, Education, and Welfare, *op. cit.*

[28]The NIH Committee had included one nonscientist, Emmett Redford, a political scientist, in 1975. A year later a second nonscientist, Leroy Walters, an ethicist, was added.

[29]Peter Hutt to Donald Fredrickson, Department of Health, Education, and Welfare, *op. cit.*, p. 483.

[30]For discussion of strategies of opposition, see S. Schattsneider, *Semi-Sovereign People* (New York: Holt, Reinhart, and Winston, 1960).

[31]Public hearings were held by the U.S. Senate Subcommittee on Health, of the Committee on Labor and Public Welfare, on September 22, 1976 and April 22, 1975; and the U.S. House Subcommittee on Research, Science, and Technology in April, 1977. In addition, state and local legislation to curb the research has been proposed in many states, including New Jersey, New York, California, Wisconsin, Indiana, and Michigan. See Nicholas Wade, "Gene Slicing: At Grass Roots Level a Hundred Flowers Bloom," *Science*, 195 (February 11, 1977): 558–560.

[32]In a letter to the NIH, the Friends of the Earth wrote, "We have been particularly struck by the small, preliminary steps being taken to deal with genetic engineering problems, with the parallels to the nuclear power controversy, which of course received no public debate or scrutiny for the first twenty years of its existence. Both nuclear power and genetic engineering seem to be proceeding on the assumption that they must proceed, yet no public debate had been initiated on genetic engineering even now as the impetus grows." Cf. letter from Lorna Salzman to Donald Fredrickson (May 17, 1976), Department of Health, Education, and Welfare, *op. cit.*, p. 542.

[33]Cambridge Hearings, *op. cit.* Vellucci also held an outdoor Saturday morning "market" of scientific ideas in a public square where scientists with conflicting views aired them to the public.

[34]Mayor A. Vellucci, NOVA interview, *op. cit.*, p. 14.

[35]Cambridge Hearings, *op. cit.* (June 23, 1976), p. 116.

[36]Mayor A. Vellucci, NOVA interview, *op. cit.*, p. 21.

[37]Cambridge City Council Experimentation Review Board, *op. cit.*

[38]Mayor Vellucci fought the decision to the end, threatening to appoint his own committee. In the end he supported the recommendations but wrote a letter denying all personal responsibility.

[39]Statement by a participant at a workshop on citizen participation at the National Academy of Sciences Forum, March 8, 1977. This attitude reflects a general decrease in public confidence in institutions. A Harris Poll found that between 1966 and 1973, the proportion of the public expressing a great deal of confidence in the leadership of institutions declined as follows: federal executives, 41% to 19%; Congress, 42% to 29%; major companies, 55% to 29%; higher education, 61% to 44%; medicine, 72% to 57%.

[40]Stanley Cohen, *op. cit.*

[41]Paul Berg, cited in the *New York Times* (February 17, 1977): 13.

[42]Albert Wheeler, Mayor of Ann Arbor, Testimony before U.S. House of Representatives, Subcommittee on Science, Research, and Technology (May 3, 1977). For analysis of the participatory trend in the Michigan case, see Don Michael "Who Decides Who Decides: Some Dilemmas and Other Hopes," in Stephen Stitch and David Jackson, *The Recombinant DNA Debate* (Ann Arbor: University of Michigan Press, 1977).

[43]Harold Green, *op. cit.*

[44]Cambridge Experimentation Review Board, *op. cit.*

[45]James Creighton, "The Limitations and Constraints on Effective Citizen Participation," in an address to Inter-Agency Council on Citizen Participation, December 8, 1976, argued that a representative government cannot deal directly with demands for issue-by-issue accountability. Citizen participation or public involvement is an adaptive response to these demands. Similar adaptive responses are taking place in European countries where experiments attempt to bring informed public involvement to controversial technical decisions so as to "reconcile contradictions between expertise and democracy." In Sweden, the government supported study circles involving some 80,000 people who met regularly in small groups to discuss aspects of the energy program. In Austria the Ministry of Industry developed a program to publicly debate the most controversial technical dimensions of nuclear energy. The assumption in these efforts is that open discussion can help to reestablish the trust necessary for stable negotiating relationships. For discussion of these

and other experiments see Dorothy Nelkin, *Technological Decisions and Democracy: European Experiments in Public Participation* (Beverly Hills, Calif.: Sage Publications, 1977).

[46]It is usually assumed that participation increases trust in the political process. However, the small amount of data available suggests that participation is not necessarily a significant determinant of trust. A survey of voters, nonvoters, and those who participate in political meetings found little difference in their trust in government. Participation is, however, associated with a sense of political efficacy, the feeling that one can actually influence the action of governments. Yet here too, cause and effect relationships remain vague for initial expectations about efficacy may have inspired the participation in the first place. See Robert K. Yin et. al., *Citizen Organization*, RAND Corporation report for U.S. Department of Health, Education, and Welfare, R 1196 (April, 1973).

[47]See for example a proposal by the American Arbitration Association, in Donald B. Strauss, "Mediating Environmental, Energy and Economic Tradeoffs," AAAS Symposium on Environmental Mediation Cases, Denver, Colorado (February 20–25, 1977). See also K. R. Hammond and L. Edelman, "Science, Values, and Human Judgment," *Science*, 194 (October 27, 1976): 389–396.

[48]J. C. Glick, in "Reflections and Speculations on the Regulation of Molecular Genetic Research," in Lappé and Morison, *op. cit.*, pp. 189–190, proposes a profession of certified public scientists who would perform independent audits of scientific research. They would belong to a professional organization (an Institute of Certified Public Scientists) which would set standards for review.

[49]For a critique of the Science Court see D. Nelkin, "Thoughts on the Proposed Science Court," Harvard University *Newsletter on Science, Technology, and Human Values*, 18 (January, 1977): 20–31.

[50]The survey, sponsored by the National Commission for the Protection of Human Subjects of Biomedical and Behavioral Research, was done by the Survey Research Center at the Institute for Social Research, The University of Michigan, in Spring, 1976. The first report appeared October 2, 1976. For a preliminary analysis see Bradford H. Gray, "An Assessment of Institutional Review Committees in Human Experimentation," *Medical Care*, 13 (April 4, 1975): 318–328, and Bradford Gray, "The Functions of Human Experimentation Review Committees," American Psychiatric Association Paper (May 10, 1976).

[51]See discussion of the problems in regulating research in Stephen Breyer and Richard Zeckhauser, "The Regulation of Genetic Engineering," *Man and Medicine*, 1 (Autumn 1975): 1–9.

[52]Mayor Vellucci, in Cambridge Hearings, *op. cit.* p. 55.

[53]For example, the animal welfare controversy at the American Museum of Natural History in New York City began because an antivivisectionist used the Freedom of Information Act to obtain a research proposal describing the use of cats as experimental subjects. See *Science*, 194 (October 8, 1976): 162–166.

[54]James Watson, from a public speech cited in *Chemical and Engineering News* (May 30, 1977). Other labels have been culled from assorted speeches, letters, and editorials.

[55]Freeman Dyson, Testimony before the U.S. House of Representatives, Subcommittee on Science Research and Technology, (May 3, 1977). See also Phillip Handler, *Annual Report by President to National Academy of Sciences* (April 26, 1977).

[56]Note that industry is less concerned about regulation than academic scientists. They are organized to deal with it, and their main concern is that regulation protects proprietory information. See Testimony before the U.S. House of Representatives, Subcommittee on Science, Research and Technology (May 3, 1977).

[57]The Genetic Society of America claimed it took no official position on political questions and instead asked forty of its most eminent members to respond. Letters were made available through the Environmental Protection Agency.

[58]These principles have been adapted from discussions in Dean G. Pruitt, "Methods for Resolving Differences of Interests: A Theoretical Analysis," *Journal of Social Issues*, 28 (1972): 133; Iklé, *op. cit.*, chap. 7, "Rules of Accommodation," and Bryant Wedge, "Communication Analysis and Comprehensive Diplomacy," in A. Hoffman, *International Communications and the New Diplomacy* (Bloomington: Indiana University Press, 1968). The context of these sources is negotiation either in collective bargaining or in international relations.

ROBERT S. MORISON

Misgivings about Life-Extending Technologies

> Hunde, wollt ihr ewig leben?
>
> Frederick the Great to his troops

> Nulli potest secura vita contingere, qui de producenda nimis cogitat.
>
> Seneca

> I warmed both hands before the fire of life,
> It sinks, and I am ready to depart.
>
> Walter Savage Landor on his seventy-fifth birthday

THE LAST FEW YEARS have witnessed one of Western man's recurring periods of misgivings about his uniquely human ability to discover, store, and process new knowledge. Most of the current unease is more episodic than comprehensive, though it has ranged over a number of topics. The recent concern about recombinant DNA and the several conferences and reports which have developed out of this concern have at least served to focus the discussion. Several factors, however, render this case less than ideal for analysis. In the first place, it is too easy to confuse concern for the immediate dangers of the experiments themselves with a longer-term concern about the later effects of the knowledge to be gained from the experiments. Second, the intriguing nature of the work gives almost endless opportunities for the discussion to lose itself in debate about technicalities. Finally, the long-term effects are highly inferential, if not simply speculative.

On the other hand, purely abstract discussion of the principles involved in promoting or limiting knowledge tends to leave practical men vaguely unsatisfied. Actually, such philosophical approaches are still rather sparse and incomplete, in contrast to the richness of discussion involving other aspects of academic freedom (cf. Metzger this issue). As somewhat of an exception, Professor Hans Jonas has recently provided a pragmatic justification for controlling or limiting biological research on the grounds that in this area, application follows so closely on the heels of thought that the ancient immunities granted to freedom of thought cannot be extended to freedom of investigation.[1] On the other hand, Carl Cohen makes a purely logical argument that finds no syllogistic support for attempts to control scientific investigation in general or inquiry into recombinant DNA in particular.[2]

The present paper descends to a lower level for a look at another practical example, less distracting perhaps than recombinant DNA and easier to relate to current methods of decision-making in science policy. A convenient starting place is provided by one of Professor Robert Sinsheimer's sample lists of investigations that he would prefer not to see carried out.[3] We shall begin with a few remarks about each one and then turn to a more extended investigation of research on aging and the implied development of life-extending technologies.

Some Examples of Unwanted Knowledge

With one possible exception, Sinsheimer's four examples of things he would rather not know about are in the how-to-do category, rather than pure knowledge, thus, perhaps confirming Jonas's view that in the biological area it is not easy to distinguish between thought and action.

INTERSTELLAR COMMUNICATION

The purest case put forward by Sinsheimer concerns interstellar communication. He feels that current attempts to communicate with other intelligent beings elsewhere in the universe would reveal to us the presence of organisms with greater intelligence and finer qualities than we have ourselves. This in turn might be very bad, even devastating, for our egos.

Viewed from the standpoint of practical social decision-making, this concern seems exaggerated. The number of scientists who are really interested in the problem is actually quite small. I have known two of them personally, and although I have been charmed by their enthusiasm and the undoubted ingenuity of their approaches, it seems very unlikely that they can command the interest of a sufficiently large section even of the scientific community, let alone the tax-paying public, to assemble the resources necessary for a comprehensive technical approach.[4] It now seems clear, for example, that knowing what they do now, the American people would probably not have followed President Kennedy's decision to put men on the moon; and the business of aiming a multitude of radio or laser beams out into space in the hope of receiving some intelligent reply several decades from now would appear to have even less public appeal. On the other hand, there is already a good deal of listening in the ultrahigh frequency radio bands with very high gain antennae. Most of this is directed at assessing the inorganic properties of the cosmos such as black holes, pulsars, quasars, and the like, but from time to time some individual investigators may intentionally scan for patterns which might occur as a result of thinking and planning by other beings. In any case, such patterned radiation would be quickly recognized if it were encountered simply incidentally in the course of listening undertaken for other purposes.

In other words, it turns out to be difficult, if not impossible, to launch a meaningful campaign against interstellar communication with intelligent beings for two rather different reasons. In the first place there is no very serious move to launch a large-scale effort of this kind, so no obvious target to shoot at. In the second place, there is no likelihood of a successful effort to stop conventional radio-astronomical observation simply because it might quite incidentally reveal the existence of beings much brighter than we are. The information it has provided and the cosmological theories that have followed are far too exciting and important to be put off by the remote possibility that the observations might also reveal the existence of better creatures.

The chances of such incidental discoveries are small, but they are not zero. At one time, in fact, the regular and surprisingly rapid pulsations of the objects now called pulsars raised the possibility of generation by intelligent beings. Other explanations quickly proved to be more likely, but pulsations in some equally regular but more intricate pattern—the ratio pi, or some other dimensionless but significant number—would be much harder to explain without invoking intelligent intention.

In summary then, as a matter of practical politics, it would seem unlikely that a campaign to prevent the possible discovery of beings better and/or brighter than we are could be undertaken. There is little in the way of an organized effort to oppose, and the chance of incidental discovery, though finite, is so small that it is overwhelmed by the excitement and interest of the other information being revealed almost every day by the same technologies. Finally there is by no means a unanimous public opinion on the deleterious effects of learning about other inhabitants, just as 500 years ago there was no complete consensus against revealing that man did not occupy the center of the cosmos and was probably not created by God in His own image. Indeed Sinsheimer's contention that the meeting of two cultures is always bad for the "inferior" one is not necessarily so obvious as he implies. Most of us who are descended from the Gauls, the Picts, and the Celts of Northern Europe, for example, seem to regard their contact with Roman civilization as a net gain. Similarly, Japanese scholars acknowledge a long-standing debt to early contacts with a more technically advanced China. Furthermore, although the contact between the New Bedford whalers and the Polynesian culture of the Pacific is painful to remember, it is scarcely paradigmatic. For one thing, it is not so easy as it once was to be sure which side was the most "advanced." Thus, even though there is no widespread enthusiasm for a large effort directed at interstellar communication, it is not unreasonable to suppose that a majority of the people are at least mildly pro rather than vigorously con. Finally, there may be a lingering strain of Calvinism that leaves some of us with what is, perhaps, a not wholly examined faith that it is best to know the truth about ourselves even if it places us a little (or even a lot) lower than the angels.

SEX DETERMINATION

Sinsheimer's fears about a quite different search for a technology to determine the birth of children of a particular sex fail to be supported on almost the opposite grounds. Here, in fact, man has always possessed an effective technology. Indeed, in a wry moment the current doyen of human genetics recently told an international conclave of anthropologists that man first became human not when he learned to fashion stone tools, or not even when syntax emerged from the deep structure of his brain, but when he consciously recognized the need to control the increase of his population by committing female infanticide.[5] Since that time, infanticide has apparently been characteristic of most societies. In this connection it is worth recalling that the Irish historian W. E. H. Lecky counted as one of the two great merits of Christianity that it managed to stop the practice in ancient Rome (the other was the ending of the gladiatorial games).[6] Versed as he was both in the history of the ancient world and the lives of the saints, he had some difficulty in deciding whether these merits of Christianity entirely balanced the evils of the religious wars and the Inquisition which stood in such sharp contrast to the toleration of classical times. However that may be, it should be reassuring to Professor Sinsheimer and his followers to reflect that the use of the effective technology of female infanticide, helpful though it apparently was in maintaining the population of many societies at reasonable levels prior to the invasion of Christianity, seems never to have been followed by the kind of social dislocation which he fears. Indeed, the recent invention of amniocentesis, coupled with the contemporary complacency about abortion has already placed in our hands an acceptable means of sex determination[7]; but it is simply not being used for this purpose on any detectable scale. Finally, even if one grants that the current technology is too cumbersome and even too ethically suspect to command wide public acceptance, there are sound empirical reasons for doubting that an entirely convenient and morally acceptable method would in fact lead to a serious dislocation in the ratio which now exists between the two sexes.[8] Opinion pollers have looked into this topic and discovered that the distortion in the sex ratio of the newly born, if any, would be rather less than that which followed the differential loss of young males in the two world wars in Germany and Russia. Again, therefore, it is not necessary to reach the question of principle. The public has apparently already decided on purely practical grounds that it would not use any foreseeable technology to cause a grave dislocation in the proportion that normally exists between the sexes.

GENETIC ENGINEERING

Others are arguing the case for and against research on recombinant DNA and by extension, research on other types of "genetic engineering"; and the jury is not yet in. In any case, it is difficult to predict the reaction of the public at this time, since the probable technologies are likely to be somewhat cumbersome, at least at first, and the results not fully reliable. The explanation of the widespread public anxiety about the possibility of altering genetic material may be looked for in a mixture of fear of the unknown with a deep sense that one's personal identity is closely dependent on one's genetic constitution rather than on any substantial showing that there will be immediate widespread use of the new methods as they become available. Who, after all, is going to want to have a child by the cumbersome artificial process known as cloning? Nobody, I should think, except for a few families who could not be reasonably sure of having a more or less normal child in any other way. Similarly, we have already noted the survey results which show that surprisingly few parents would make use of even a very simple and convenient method for sex determination if such were to become available. It is somewhat more difficult to predict the use of more complex forms of genetic engineering. However, the only capability which would seem immediately attractive is the substitution of normal genes for those which under current conditions may be regarded as almost wholly deleterious—sickle cell anemia or hereditary cholesterolemia, for example—and these affect only small fractions of the population. Much more complex choices will confront us when we think we know enough to tackle polygenic traits which may have important expression in emotional behavior or intellectual capacity. However, such possibilities lie so far in the future and are so full of uncertainties as to preclude serious consideration at this time. Those who oppose the development of these technologies frequently refer to the probability that some malevolent dictator would use them to turn out endless replicas of a human design reflecting the particular needs of his form of totalitarian state. But it is impossible to see how anyone could do this unless the state had already reduced its subjects to a condition of abject compliance. In other words, large-scale use of genetic engineering to implement the desires of the state could only be a result, not a contributing cause of repression.

Professor Sinsheimer's fourth example, that of research into the aging process, provides a much more appealing case for analysis at this time. Indeed, it should perhaps be made clear at the outset that I happen to share Professor Sinsheimer's view that if research into aging actually results in an important extension of the normal life span, it will indeed lead to serious social and personal dislocations which we might prefer not to experience. Let us then look at the high points of research into the aging process, with a view to seeing what might be done in practice to limit the growth of such knowledge if it were thought desirable to do so.

Aging Research and the Possible Effects of Life-Extending Technologies

First, and necessarily very briefly, what is the biological situation? As everyone knows, what the biostatistician refers to as life expectancy has gone spectacularly upwards in advanced countries during the last hundred years. This does not mean that the very old are any older than they used to be. It is simply that a much higher proportion of people actually reach old age. This has come about for a variety of reasons, ranging all the way from more and better food, more and better housing, to better sanitation, and, finally, "miracle" drugs. Fewer and fewer people fall prematurely to infectious disease and the other killers of young people. Instead of the few percent that once reached the psalmist's 70 years, now nearly two-thirds reach that venerable point. But by 100, essentially everyone has passed on, just as they always did. Presumably there can be no serious misgivings about further research to enable a larger number of people to avoid the hazards of early life so that they can stay the normal course. We have, in fact, done most of that job already. Thus we are also already familiar with the resulting problems, which center, of course, on the greater proportion of people who must be supported after retiring somewhere between the ages of 60 and 75. Any further changes that may occur as a result of continuing to whittle away at life-shortening diseases pose problems of degree only.

The possibilities that lead to uneasiness about research on life-extending technologies are those which might extend the so-called normal life span so that a substantial fraction of the human race might live several decades longer than they do now. Some enthusiasts even talk about life without end. Since these matters have not been the subject of widespread publicity and discussion, it may be well to devote a few paragraphs to outlining what appears to be the present biological situation. Although a good many philosophers and some biologists enjoy arguing about whether or not aging is a disease[9] and death a natural or an unnatural event,[10] the evidence comes down strongly on the side of aging being (at least in complex organisms) an entirely natural process, depending on specific mechanisms that have evolved to ensure that in a given species death will occur within a specific age range. Indeed, death is probably essential to the evolutionary process as we understand it. For example, one of the most useful definitions of Darwinian fitness revolves around the possibility of living long enough to have a given number of offspring. Those who do not survive to the reproductive period or who produce only a very few offspring are regarded as less fit than those who live longer and have more. Clearly, if all animals and plants were immortal, the concept of Darwinian fitness would lose its meaning and in the long run there would actually be little room for change. The cultural equivalent of death in this sense is, of course, the enforced retirement age, without which assistant professors could not be promoted and new ideas would find their way into the academic community with even greater difficulty than they do now. So much for the theoretical justification for regarding death as a natural event.

STATUS OF CURRENT RESEARCH ON AGING

Recent research into the actual processes of aging now suggests quite strongly that not one but several processes are at play in determining the rise and fall of the seven ages of man.[11] Aging and death are not the results simply of wear and tear, but of what some critics of modern industrial practice might define as planned obsolescence.

Perhaps one should pause at this point to observe that research on aging has never until recently attracted the serious interest of a significant fraction of the biological community, nothing comparable, for example, to the effort in genetics or the processes of early development and differentiation. Within the last decade, however, several leads have been opened that have greatly improved the outlook for a workable understanding of aging processes. The lines of inquiry may be roughly distinguished as follows: (1) the nutritional and metabolic, (2) the control of cell proliferation and differentiation, (3) the immunological approach, (4) changes in connective and supporting tissues, (5) central nervous control of the endocrine apparatus. It seems highly probable that these approaches are not to be regarded as representing independent or alternative explanations. On the contrary, the different mechanisms may well work together in a coordinated way to produce the end result, with which we are only too familiar. Indeed, the multiplicity of mechanisms is probably another illustration of how Nature manages to build a high degree of redundancy into the machinery for carrying out functions that she regards as particularly important.

The nutritional and metabolic theories had their debut in a series of experiments undertaken many years ago by Professor Clive McCay at Cornell in which he showed that rats kept on a diet marginally deficient in total calories tended to live over fifty percent longer than those on a normal diet. The growth and development period as well as the duration of adulthood tended to be proportionately expanded. So far there is no certain explanation, but it appears that the response may have been mediated through the same central nervous and endocrine control mechanisms that are under current study. Somewhat similar results may be obtained by maintaining animals at lower than normal body temperatures, but, at present, neither of these approaches offers encouraging leads for early application.

The modern tendency to explain large events in terms of small units has directed attention to the fact that cultures of normal cells taken from higher organisms go through a definitely limited number of divisions and then cease to grow and begin to wither away. This is not so true of malignant cells, and there are cancer cultures which appear to be immortal, at least in principle. This suggests strongly that the cessation of growth in normal cultures is a deliberately controlled process; and it directs attention to the possibility that there may be genes for growth, differentiation, and cell division that can be turned on and off by some as yet unidentified process. Indeed, the attention of many forward-looking biologists has recently been turned from the action of genes themselves to that of the relatively unknown substances responsible for the activation and deactivation of genes at various stages in the process of growth, differentiation, development, and decline. At this stage it may be anybody's guess as to when we will understand these processes well enough to be able to control at least some of them ourselves.

Other, less well-understood, perhaps incidental factors may influence the process of aging at the cellular level. There is some evidence, for example, that an accumulation of oxidants may lead to an excessive production of "free radicals" or unsatisfied portions of molecules which, in turn, may hasten the aging process by attacking and degrading normal cellular constituents. The theory, though unsubstantiated, is an interesting one because it leads to an immediate possibility of doing something to slow the process by treatment with antioxidants. Some investigators are convinced enough by this chain of reasoning to have started ingesting fairly large amounts of vitamin E every day.

Other investigators draw attention to the progressive decline in the activity of the body's immune apparatus. Not only does this make the aged more susceptible to ordinary infectious disease, but it also slows down the elimination of abnormal molecules which may occur in the process of normal metabolism and thus might allow the appearance of malignant tumors which would normally be eliminated by the tissue immune systems. Even more dangerous perhaps is the fact that as the immunity towards foreign substances decreases, the tendency to become immune against one's own normal tissues may increase. This phenomenon of so-called autoimmunity may be a major factor in the development of atherosclerosis and other degenerative diseases. All this is still rather speculative, but of a good deal of potential consequence, since we already know how to stimulate or turn off some related kinds of immune reactions.

Finally, there are the theories of aging that emphasize various central processes and the linkage between the hypothalamus of the brain, the pituitary gland, and the rest of the endocrine system. Few things are more dramatic, of course, than the physical and psychological changes that accompany adolescence and are initiated in some way by this same neuroendocrine axis. It requires no great imagination to hypothesize that the slowing down of the body's ability to meet stresses or undertake great enterprises may be due to a program that deliberately reverses the exciting events of puberty.

This limited sample of the work now picking up speed in the general area of gerontology is by no means definitive or complete. It is submitted merely to demonstrate that there is a real possibility that something will happen during the lifetime of those now living to extend their lives by perhaps a decade, or perhaps even two or three decades. The probability is sufficiently great to remove consideration of the possible sociological and psychological repercussions from the category of idle speculation to that of serious technology assessment. It also gives weight to the concerns of those who wonder if it will really be nice to know how to expand the life span on a large scale.

Decision-Making and the Public Interest

We turn now to a consideration of these events in the light of the social machinery for the direction or limitation of further scientific investigation. The situation is different in several ways from that which led to the current excitement about limiting research on recombinant DNA. In the first place, there has, until recently, been very little research on aging as such. Most of the hypotheses and investigations which we have just summarized have grown out of work originally undertaken with quite different ends in view. Modern investigations of immunity may be said to have begun with Pasteur and his many followers in the area of infectious disease. More recently, immunologists have been concerned with the more esoteric problems of autoimmunity and tissue rejection as related to the problem of organ transplants. The whole matter of growing cells in tissue culture has been of technical interest since Ross Harrison first succeeded in growing nerve cells in test tubes shortly after the beginning of this century. It now forms a large part of the activity of that army of investigators interested in molecular genetics, normal growth and development, and cancerous degeneration. The relationship of the neuroendocrine system to aging is but one relatively small facet of the overall control of bodily function generally.

Although there has been a recognized interest in gerontology with a sort of subdepartment of interest in the biological aspects of aging for some decades, it has been a rather loose association with nothing resembling the invisible colleges which are said to dominate so much of the more recognized areas in modern biological research. On the contrary, gerontologists have on the whole been an individualistic, often idiosyncratic lot, not fully freed from the disquieting traditions of the monkey gland injectors of the early twenties. Thus, there has been no organized scientific lobby to promote research on aging, nor does the inner dynamic of the field itself have that self-energizing, self-perpetuating quality that is said by Jacques Ellul and others to take decision-making out of human hands and place it in some mystical entity associated with technology itself.

The role of the public has also been quite different from that in many other areas of science. On the whole, the public is thought to have been relatively unprepared for the atomic bomb, the population explosion, organ transplantation, or the still equivocal promises of genetic engineering. All these and many more, including presumably the automobile, are currently alleged to have been foisted on an unsuspecting, unprepared public as a result of the autonomous scientific technological complex referred to above.

In contradistinction to these classic cases, there is nothing esoteric or unfamiliar, or, on its face, undesirable, about the idea of extending the life span. Indeed, the wish for such technology has been part of the human condition since the beginning of time, and there are many elaborate mythologies concerning methodologies for living forever. In modern times, pressures to investigate the possibility of extending the life span have come as much or more from the public outside than from the inner dynamics of science. The most explicit recent move has been the bill passed in 1974, which took the investigation of aging away from the Institute of Child Health and Human Development and gave it an entirely new, independent institute of its own. It may be instructive, especially for those who tend to deplore the lack of public participation in the making of science policy, to give a brief review of the events leading to this decision. In this instance, at least, the major impetus seems to have come from lay groups quite outside either the government or the scientific community. One of the most active appears to have been led by Mrs. Florence Mahoney, an important figure in the Washington social scene with access to a number of legislators, among them Senator Thomas Eagleton, who played an important role in piloting the bill through the Senate. In this respect the situation is somewhat similar to the one that resulted for so long in rapidly increasing budgets for the national institutes devoted to cancer and heart research, where Mrs. Albert Lasker played a similar facilitatory role. In these two instances, however, several distinguished and energetic members of the scientific community added their expertise to the lobbying team. So far, no analogue of the stature of a Sidney Farber or Michael DeBakey has emerged to carry the banner of aging. The public interest was, however, more directly represented by a number of organizations and affiliations, one of the most aggressive of which was appropriately named the Gray Panthers.

The executive branch of government, including the National Institutes of Health themselves, seem to have been opposed to the bill and most government representatives called to testify took a position in opposition. This may account in part for the very slow progress made by the new institute since its inception. Indeed, it is not yet clear exactly what its program will be. The preamble to the bill speaks enthusiastically about keeping "our people as young as possible as long as possible," a phrase which certainly implies emphasis on what we have called life-extending technologies. On the other hand, a good deal of the pressure for the legislation came from those interested in ameliorating the condition of our *current* senior citizens. The appointment of a psychiatrist to head the institute suggests that at least the initial emphasis will indeed be on programs for improving the psychological and social context for the elderly in a society that increasingly emphasizes the nuclear family, day care centers and singles bars for the young, and early retirement, emotional isolation and sunset homes for the elderly. However, a biologist has been appointed as deputy director. Presumably he will devote much of his attention to making our population live "as long as possible."

Although Congress several years ago set up an Office of Technology Assessment in response to the growing recognition that much of the science and technology sponsored by government agencies can have undesirable as well as desirable effects, it does not appear that its help was sought in predicting the overall results to be expected from increasing the life span. Nor was the opposition in NIH based on any serious misgivings about the results of the research to be pursued, but rather on administrative considerations. About the time the bill passed, however, the National Science Foundation did contract with a futurology group to work out what some of the implications might be. The report is not in yet, but it is understood to deal primarily with predictions regarding the social, financial, and political stresses that might result from a series of hypothetical changes in age distribution of the population. There will also be at least one chapter on some of the value problems which may be raised. While we are waiting for the final report, it may be possible to speculate on what some of the more significant results might be, and to ask ourselves whether a clearer view of possible adverse effects might have changed the decision-making process just described.

As pointed out earlier, society is already familiar with most of the problems likely to emerge from the current ability to pull an increasing fraction of the population through the normal life span. Clearly our social security and pension arrangements are in for greater stress during the next twenty years. The ensuing fiscal difficulties, coupled with an increasing recognition of the psychological damage done to many individuals required to retire at what they regard as an early age may ultimately require some revision of present rigid retirement rules. But this can happen only after a lively debate with the younger generation, themselves torn between a desire for upward mobility and a desire to free themselves from the burden of an increasing number of nonproductive oldsters.[12]

But there is no reason to suppose that such problems are insoluble. At quite a different level, but perhaps with even greater success, society is already facing up to dealing with those life-extending technologies that are capable of extending the physical signs without the psychological and social concomitants of life. Although the marginal cases are still capable of stimulating lively debate among those with dogmatic views on both sides, there seems a growing consensus, in principle, that there should be reasonably well-defined limits to the use of what are sometimes referred to as extraordinary means for prolonging life. The right of the individual to refuse treatment so long as he is fully competent is now universally accepted; and efforts are being made to establish the power to refuse in advance treatments that might be recommended at a time when the patient may have lost the power of expression.

In summary, though there have been strains as medicine has succeeded in bringing a larger and larger proportion of the population into the seventies and eighties, there is nothing to suggest disaster if three-quarters or more of the population reach the later decade. On the contrary, the sharp reductions already achieved in what might be called premature death have been welcomed by almost everyone, and there is little or no opposition to continuing the search for new remedies and life-extending technologies in this sphere.

Those who would like to slow down or even stop research into the aging process and the development of life-extending technologies must be thinking of some combination of new procedures that would enable substantial portions of the population to reach unheard of ages. For example, in his recent book, significantly entitled *Prolongevity*, the respected science writer Alfred Rosenfeld predicts "that by the year 2025—if research proceeds at reasonable speed—most of the major mysteries of the aging process will have been solved and the solutions adopted as part of conventional bio-medical knowledge; and that some of the solution will by then already have come into practical use to stave off the ravages of senescence."[11] From there he goes on to talk about possible consequences and, without making specific predictions, seems to assume that the possibilities are virtually unlimited. It is considerations such as these that lead to the kind of misgiving expressed by Professor Sinsheimer and other thoughtful people.

Possible Social Problems

The most obvious problems likely to follow an unexamined extension of the life span are primarily social in nature. First, and most obviously, if all people live to be much older than they are now, the population of the world will increase proportionately and the ratio of the various age classes will be seriously distorted. The same sort of thing is true, of course, of the disease control measures already in train, but in advanced countries we have already experienced most of the increase in population and most of the change in distribution among age classes that is likely to occur from this source. A hundred years ago, for example, only a tiny percentage of the population was over 65. Now the figure is nearly 10 percent. If everybody lived till 80, and birth and death rates reached equilibrium, only about 19 percent would be above 65. But if everybody lived to be 130, half the population would be beyond what we now regard as the retirement age. Of course it will be contended that many of the proposals for extending the life span would also extend the period of useful and enjoyable life, so that the best part of life would be stretched out the longest. Indeed this is the apparent intention of the recent legislation for "keeping our people as young as possible as long as possible." Even were this to prove to be the pattern, and it is by no means certain that it would, there is no doubt that the stretching-out process would cause stresses and strains in society, as retirement ages were forced up and younger individuals found it harder and harder to rise to positions of responsibility and eminence. Perhaps all these practical problems could be solved by good will and careful thought, but the stress on existing social institutions is sure to be orders of magnitude greater than that caused by any of the life-extending technologies which have been available so far.

Distributive Justice

Even if these difficulties could be overcome, however, we are still faced with interesting philosophical and biological questions. The philosophical question is perhaps most easily exposed within a utilitarian framework. Granting that there must be some optimum population for the world, an increase in longevity requires that fewer people be born in a given time. This means that fewer people would have the opportunity to experience the joys and sorrows of life, while others simply experience more of the same. If one applies the doctrine of diminishing returns to the situation, the total amount of happiness would appear to be reduced. Furthermore, even the current enjoyments of old age might lose some of their remaining piquancy if one knew that they were likely to continue almost indefinitely. The days would no longer dwindle down to a precious few, but on the contrary would stretch out into the infinite boredom of doubly diminishing returns. The sum of these effects on individuals would in turn reduce the total sum of human happiness.

So much for the strictly utilitarian aspects. From the biological points of view, the situation is, if anything, even more questionable. Among biologists, as we have noted, the presumptive explanation for the species specificity of aging and death is closely related to the mechanism of evolution. The idea, in its simplest terms, is that one is programmed to live long enough to demonstrate one's fitness and then disappear to make room for a new model. Actually, of course, human biological evolution proceeds very slowly and human "fitness" cannot be described simply in terms of reproductive capacity. Postmenopausal women and men do have contributions to make to the survival and further development of the race, and anthropologists and sociologists are expanding the fitness concept to include kinship groups. Nevertheless, the principle that the obsolete should make room for the new should ultimately apply to both cultural and biological evolution unless something quite remarkable and unforeseen is built into the new life-extending technologies.

The people whose lives are prolonged are likely on the average to cherish differentially the various ideas and practices they and others of their generation have developed. Inevitably the process of social evolution will be slowed down, simply, so far as one can see, for the selfish advantage of that tiny fraction of humanity that happens to be alive at the time the life-extending technologies become available. Although it is probably impossible to prove any ethical proposition beyond a peradventure, the wish to live more or less forever at the expense of the opportunity of an indefinite number of potential others to live at all, appears on its face to be unjust. Furthermore if the biologist's belief in the importance of change and the encouragement of what might be termed new experiments in living can be extended to apply to the human condition, there is perhaps a kind of cosmic effrontery in wanting to live forever.[12]

Individual Psychological Effects

In addition to the social arguments, which would seem to be compelling, it is worth considering the situation from the point of view of the individual whose life is to be extended. All of us know in a general way that we must die sometime, and virtually all of our life-planning takes this contingency into account in some way or another. Indeed, it is commonly observed that many men consciously risk their lives in order that the recognition of death may emphasize for them the uniqueness and importance of life. Technologies which would postpone death to some unidentifiable point in the future might result in a life of constantly increasing boredom so that the temptation to incur deliberate risk would be constantly increased. Finally, the victim of immortality might find himself entirely preoccupied with deciding when and how to commit suicide. Not everyone is likely to feel this way, of course, but the probability that some will is high enough to require planners for immortality to plan for this further contingency.

Another set of considerations arises out of the fact that in most lives the balance of pain and pleasure is a somewhat precarious one. One of the consolations for the pain that man has invented for himself is the reflection that some things can't be changed. What's done is done, and we had better learn to put up with it, or perhaps even put it out of our minds. One of the reinforcements for this point of view is the reflection that life is short and does not provide us with many opportunities for correcting mistakes. A life of longer or infinite duration, however, might take away this consolation; with so much time at our disposal we would have no excuse for not going back to fill the gap or correct the error. Thus the total burden of guilt and depression is likely to increase.

Perhaps it is worth nothing also, although it would take a good deal of further research to establish the point beyond a doubt, that most of the things men set their hearts upon ultimately decline and pass away. It may be fortunate for most of us that we also decline and pass away before the homesteads we cleared, the families we sired, the looms we invented, the clipper ships we commanded, and the empires we were determined not to dissolve reached their peaks and began their inevitable declines.

One final speculation may be worth listing for future exploration although it can be put forward only very tentatively and largely on the basis of anecdotal and literary evidence at present. It has often been asserted that in order to live at all, human beings need some structure of myth and illusion to give meaning to their lives. It is also at least part of conventional wisdom to assert that illusions tend to disappear over time. Old men, by and large, and overlooking the deathbed confession, seem to have fewer illusions than young ones. There is thus a distinct possibility that prolongation of life will mean a steady decline in the number of illusions for at least a substantial though admittedly unknown fraction of the human race. We thus may have to face the paradox that the ultimate result of life-extension technologies might be an increasing recognition of the pointlessness of the entire exercise.

What all this adds up to may be simply this. Ever since man became a conscious, thinking, designing human being, he has molded his individual philosophy and cultural institutions to the biological fact that we all must die. In a thousand different ways, some conscious and some deeply embedded in their unconscious, men recognize the cyclical nature of their lives and the fact that all things have their seasons. Our most basic ideas about the meaning of life are based upon what we know and feel about death. So fundamental are these assumptions, convictions and feelings that it would be presumptuous for any research group to believe that it could detach itself sufficiently to make anything like a technology assessment of the psychological and social impact of any substantial lengthening of the normal life span, especially one which leaves open the possibility of infinite extension. All one can reasonably say at this time is that the impacts and readjustments will be profound and it might be wise to be thinking about them before they happen. In this we must all sympathize if not completely agree with Professor Sinsheimer's misgivings, but are we equally drawn to conclude that the best way of coping with the problem is to cut off support for all research with a possible bearing on the aging process? In preparing to answer these questions we might first ask about the practical or political possibilities.

How Should Society Prepare Itself?

The love of life is so thoroughly built into us that at first glance it seems impossible for any substantial fraction of the human race to want anything else but its indefinite extension. Indeed, the momentum in that direction is so great that it would probably be impossible to slow research on life-extension to any significant degree. Until recently, the science of gerontology has languished, not because people were not interested in living longer, but because the weight of scientific opinion was against the possibility that much could be done about it. The recent uncovering of several promising leads has changed all that. Gerontologists are going about with smiling faces, and appropriations are beginning to increase. Only now and then does one hear a suggestion that aging should be put on a list of those things we ought not to try to know more about. The debate on whether or not we can save ourselves from ourselves simply by trading enforced ignorance for our tradition of free inquiry is only now beginning (cf. References 1–3), but it seems very unlikely that many of our larger problems will be solved in this way.

Thus the move to accelerate research into the aging process is gaining momentum from both within and without the community of biological scientists doing work relevant to the problem. It is doubtful then that, even if they wanted to, any small group of thoughtful and responsible scientists could slow down the movement. It is almost equally hard to see how even the most fanatical leaders of Science for the People could conjure up the same kinds of horrible imaginings about aging research as they have raised in the case of recombinant DNA and other aspects of "genetic engineering." Paradoxically, the long-term results of lengthening the life span are quite likely to have far more deleterious impacts on society than either the immediate or long-term effects of the raising of "test tube babies," but they do not lend themselves nearly so well to the arts of propaganda.

We have already given reasons for doubting that genetic engineering is likely to be widely used for undesirable purposes in the foreseeable future. On the other hand, it is hard to think of any new biological technology that is likely to be adopted so quickly and so widely as one that would make us feel as young as possible for as long as possible. Consider, for example, just one of the likeliest candidates for early availability—a pill which would slow the normal tendency to form cross-linkages between adjacent collagen molecules which, among other things, produce the hardening of the arteries, the wrinkling of the skin and the stiffness of the joints that make life less comfortable even before retirement and may end by making it scarcely worth living. Would even the Elluls and Roszaks, much less the Sinsheimers or Chargaffs, turn their wrinkling cheeks away from such a proposal?

Our main point is that the very universality of appeal that makes life-extension so patently disruptive and amply justifies the concerns so far expressed, at the same time renders it immune from social control by limiting research on the aging process or banning the development of the consequent technologies. The good results are obvious and immediate and fit neatly into the long-standing myths of man's fondest hopes. The unhappy "side effects", on the contrary, lurk mostly in the future, many of them shrouded in a kind of actuarial obscurity. As a practical matter it is simply very unlikely that any congressional committee or regulating board would elect to slow down the research so full of immediate promise. In cases such as this we simply do not reach a question of principle in which a traditional right to knowledge or freedom of inquiry is pitted squarely against some hypothetical right to remain ignorant. The reality is that there are simply too many good, or at least desirable, results to be expected from most of the inquiries into aging that can be easily foreseen. If there are other less desirable results, they will have to be dealt with *one by one as best we can*, just as we are slowly learning how to deal with the undesirable results of other apparently attractive technologies.

It is not the purpose of this paper to suggest specific solutions to the numerous social and personal problems that will probably attend success in lengthening the life span. This is clearly a task for the future. In a time of diminishing faith in knowledge as such, and in a collection of papers dedicated to examining its possible limits, it seems not out of place to note that the advent of new knowledge has always required the discovery of still more knowledge to ensure its use for the benefit rather than the destruction of mankind. On balance, as even Professor Sinsheimer concedes, the results have been benign. Today's problem is one of making sure that the bottom line still remains in our favor. It may well be that the field of science policy is one of the last great areas of human affairs to cling to a belief in an invisible hand that will guide the effort in a benign direction so long as each individual research worker is allowed to pick his own problems and do with them exactly as he likes. Indeed the majority of science workers are probably ready to agree that this is too simple a faith and that social institutes "like Cal Tech have an obligation to be concerned about the likely consequences of the research they foster" (Sinsheimer, this issue)—and by implication to plan their expenditures in a way as to maximize the good and minimize the evil consequences. But it seems very doubtful that the best way to begin the planning exercise is to search out and eliminate a series of basic research projects now in progress that might produce knowledge that could have undesirable as well as the expected desirable effects at some unspecified future date. The infant art of technology assessment was in part invented to avoid just such summary executions in advance of the crime.

References

[1]Hans Jonas, "Freedom of Scientific Inquiry and the Public Interest," *Hastings Center Report*, (August, 1976): 15–17.

[2]Carl Cohen, *When May Research Be Stopped*? (to be published).

[3]Robert L. Sinsheimer, "Recombinant DNA: A Critic Questions the Right to Free Inquiry," *Science*, 194, (October 15, 1976): 303–306.

[4]For a popular account of the technology and plans to ask Congress for a "modest but full-time search for artificial radio signals from beyond our solar system" see Timothy Ferris, *New York Times Magazine* (October 23, 1977); 31.

[5]James Van G. Neel in *Proceedings VIII International Congress of Anthropological and Ethnological Sciences*, vol. 1 (Tokyo: Science Council of Japan, 1969, pp. 356–361 reprinted in *Science*, 170 (November 20, 1970): 1815–1822.

[6]W. E. H. Lecky, *History of European Morals from Augustus to Charlemagne*, vol. 1 (New York: Appleton, 1860).

[7]The procedure known as amniocentesis involves withdrawing a small amount of amniotic fluid containing fetal cells. The presence or absence in these cells of "sex chromatin" identifies the fetus as either female or male at a stage early enough to allow an induced abortion of a nonviable fetus of the undesired sex.

[8]Charles F. Westoff and Ronald R. Rindfuss, "Sex Preselection in the United States: Some Indications," *Science*, 184 (May 10, 1974): 633.

[9]H. Tristram Engelhardt, Jr., Ph.D., M.D., "Is Dying a Disease?" *Values in Life Extending Technologies*, a report to the National Science Foundation by the Institute of Society Ethics and the Life Sciences (1977).

[10]Peter Steinfels and Robert M. Veatch (eds.), *Death Inside Out* (New York: Harper & Row, 1975); see particularly the essay by Ivan Illich, pp. 25–42, and the four essays on "The Naturalness of Death" by Paul Ramsey, Robert S. Morison, Leon R. Kass, and H. Tristram Engelhardt, Jr.

[11]Much of the material in this section will be found conveniently if a little too attractively packaged in Alfred Rosenfeld, *Prolongevity* (New York: Alfred A. Knopf, 1976).

[12]As this goes to press the debate has burst into public notice with the passage of the House bill eliminating some retirement ages and raising others. The ultimate outcome is, however, still unclear.

[13]This paragraph is obviously subject to the criticism that it is written from a parochial Western point of view clearly contaminated with notions of progress. It may also be uncomfortably close to confusing a biological "is" with a human "ought." A more sophisticated presentation might simply note that an extension of the life span would highlight the need for *some* decision on these issues.

GERALD HOLTON

From the Endless Frontier to the Ideology of Limits

THE PUBLIC DISCUSSION in the United States about constraints on scientific research seems to have come upon us with startling suddenness. Ironically, it surfaced at the very moment when scientists have the right to think that the fundamental advances being made are better than ever, and when scientists are loudly asked to help solve the vast problems in such areas as energy and the environment. Scientists find themselves rather bewildered. For decades, they have proudly accepted P. W. Bridgman's well-known operational definition, "The scientific method is doing one's damndest, no holds barred." Now, they are asked to add the phrase, "—except as laid down in guidelines issued in Washington and by the local town fathers." The old image of science as the "endless frontier," on which a whole generation has been brought up, seems to be giving way in some quarters to the notion of science as the suspected frontier.

Many observers date these changes from the first public expressions of concern by biologists about the possible side-effects of doing research on recombinant-DNA molecules. If the debate is of such recent origin, can it last? In a country where even the most sensational and preoccupying activities can suddenly disappear in silence, are we really dealing here with a serious challenge, rather than merely a highly-visible but short-lived excitement?

The answer seems to be a clear "yes." In this epilogue to a handsome set of essays I shall sketch reasons for believing that we have only begun to struggle with such problems. For whether one likes it or not, the disputes concerning the wisdom or danger of placing "limits on scientific inquiry" may have been inevitable and were perhaps overdue. Depending on the specific cases that clamor for attention, the intensity of the discussions may wax or wane; but they have a certain preordained character, and in maturing form will remain with us for a long time to come.

There are more than half a dozen chief factors determining these events today, each of which can be expected to continue to exert its force in the future. I shall now take up each one in turn.

The New Visibility

The simplest of the reasons for the expected endurance of the issue has to do with the attainment by the scientific establishment of a kind of critical size. The increase in the scale of scientific and technological activity in the industrialized countries has made for visibility, and with that—as should be the case, at least in a democracy—has come a re-examination of the mechanisms of accountability. The annual cost of basic research in the United States is now close to $5 billion (two-thirds of it supplied by the federal government) and a total of nearly $40 billion is being spent on the national research and development effort (including defense and space projects, which, for better or worse, the public identifies closely with science). Clearly, any enterprise that employs on the order of one million scientists and engineers, and commands 15 percent of the relatively controllable portion of the federal budget, must be subject to mechanisms of accountability with respect to its performance and justification, in terms that taxpayers or their representatives can appreciate.

This rising level of activity, and hence of the power of science and engineering to change our world, acts in two quite opposite ways on public perception. Along with splendid new discoveries and some welcome gadgetry, there is also a higher level of risk, compared with the level of risk in the past, in periods of slower technological change. Patent dangers, outright abuses, and significant mistakes may constitute only a small proportion of on-going work; but when the rate of change of scientific knowledge and of technological advance is rapid enough, that small proportion does add up, in absolute terms, to form a visible set of cases. The public alarm system seems to be set at absolute levels, not by relative measures.

Moreover, as the rate of change is increased, it has the unsettling effect of decreasing the time interval during which the change being imposed on society can be monitored, evaluated, or even absorbed intellectually by the public—just when this public has learned to insist more and more on having some role in consequential decisions. To illustrate: in 1976 the lay citizens' group appointed by the City Manager of Cambridge, Massachusetts, to advise the City Council on guidelines for the use of recombinant DNA-molecule technology declared on page one of its generally thoughtful report:

> The social and ethical implications of genetic research must receive the broadest possible dialogue in our society. That dialogue should address the issue *whether all knowledge is worth pursuing.* [Emphasis added.] It should examine whether any particular route to knowledge threatens to transgress upon our precious human liberties. It should raise the issue of technology assessment in relation to long-range hazards to our natural and social ecology. Finally, a national dialogue is needed to determine how such policy decisions are resolved in the framework of participatory democracy.

One may try to dismiss this as a pastiche of clichés. Back in the minds of such persons is the fact that the implicit promise of science has been a rosy world free of disease, with new industries and new jobs waiting for us, a world giving us secure peace, and perhaps even the joy of understanding what those scientists are discovering in their noble quest. Little of that has come to pass. On the contrary, the very solutions proposed, such as nuclear reactors to supplement our shrinking energy supplies, now loom for many as threats to health and peace. As will be stressed in more detail further on, there is a widespread disillusionment with the explicit or implicit promises of technological solutions to social problems. And even those who were not attracted to the activist image of science and technology, who preferred the older idea of the scientists as lonely thinkers and the engineers as inspired tinkers—occasionally very useful people but essentially harmless—have had to readjust to the much greater and confusing variety of the current roles that scientists and engineers play, as employees of universities, industries, consulting firms, government laboratories and "think-tanks," or as advisors to state government, the Congress, or the President. If science ever was a charismatic profession dominated by abstract spirits, those days are gone forever.

Short of suffering crippling decreases in financial support, work in science and engineering will not be less visible and less watched in the future. It is, however, reasonable to hope that the fraction of abuses, mistakes, surprises, and other alarming problems will drop as the professionals involved become more and more sensitized to the possibility of such problems. That may be one good lesson drawn from the list of horrors of recent years, including the side-effects of pharmaceuticals and food additives, the dubious mathematics of risk-calculations, and, alas, the estimate that at least one-quarter of the world's scientists and engineers is engaged fairly directly in exploiting the "pure" advances for weapons-related research. The celebration of the basic intellectual triumphs of science has had hard competition.

The Old Credo

If the newly visible risks are a part of the public's perception about science that is not easily changed, the current self-perception of scientists may be only slightly more flexible. While today there is more diversity of views in what is lightly called the scientific community than has been the case since the 1930s, the largest proportion of scientific professionals is watching these debates with considerable apprehension. Many scientists believe that the integrity of their activity must be strenuously protected against what they perceive to be potentially serious threats from an unchecked, or perhaps ill-informed, limits-to-inquiry movement. Whether they originate from fellow scientists or other citizens, calls for any explicit limits go against long-standing traditions of academic and research freedoms as still understood by most scientists. They contradict the predominant philosophical base of science as an infinitely open system in which it is not to be feared that one will run up against questions that are, in principle, unanswerable, not to speak of questions that are unaskable. They are also contrary to the psychobiographic drive of most of the younger scientists and the reward system that has shaped the older ones. They clash violently with the world view of scientists brought up on the notion that science and optimism are virtually synonymous, that somehow the findings of science will be for the good of mankind.

One recalls Jacques Monod's remark in *Chance and Necessity* that the ethic of knowledge is the commitment to the scientific exploration of nature. The limits movement conflicts basically with such commitments; moreover, it comes, ironically, just at a time when the results of scientific and technological work, both in terms of quantity and in terms of quality, contain more examples of ingenuity and beauty by any standard than ever before.

Some of the most successful and visible scientists, when calling for as unrestrained an autonomy as is compatible with safety, tend to regard the very discussion of proposals for "limits" as dangerous and self-fulfilling. The period of accommodation, on either side, therefore, is likely to be a long one, the more so when scientists find themselves confronted with extreme proposals, or with banners such as the opening to a recent draft statement from the World Council of Churches: "However robust, the faith of yesterday in the power of science and technology is today misplaced and therefore misleading. In fact, the reductionist, triumphalist, manipulative approach of science and technology has itself been in large part responsible for our predicament."

In fact, however, scientists or engineers today are on the whole far more ready to take care that the ethos and practice of science include protective limits and constraints than is popularly recognized. A number of internal and external constraints have been found functional and even essential (except that the red tape associated with certain regulations is threatening to get out of hand). Thus, to cite some controls internal to the practice of science, it is taken for granted that quantitative results, if obtainable, are preferred over qualitative ones; that operational definitions rather than metaphysical ones be used; that important experiments be repeated; that an attempt at relating theory and practice be made; that the work be ultimately published; that credit for priority, collaboration, and support be given; and so forth. As every modern researcher knows, getting a share of scarce resources—e.g., time, funding, and manpower to do an experiment involving one of the large machines—or getting the approval of the local committee charged with the regulation of hazardous biological agents, can lead to epic fights.

Among external mechanisms for discipline and accountability, scientists consider it natural that peer review be undertaken on funding priorities and on the soundness of procedure; that the peer review process itself be open to inspection and evaluation; that informed consent be obtained from experimental subjects; and that research involving toxic or radioactive substances, or human beings used as "animals of necessity" be placed under careful regulation; and that environmental and other impact statements be provided for large engineering constructions.

One can discern three levels of external restraints or controls on research, to which scientists tend to react with increasing alarm. Those arising from the hazardous nature of research are long-established and respected on all sides. Constraints that arise from decisions to invest in one area of research rather than another are more controversial, in good part because this has to be ultimately a political decision (made in the United States, in effect, more by the Congress and the Office of Management and Budget than by the scientists concerned). Third, controls may be imposed by nonscientists because they have qualms about scientists by themselves being able to estimate properly the hazards of a given research and/or the ethical problems involved in the procedure or its possible findings. The last is evidently the area of greatest dispute.

Until recently, most of the limits on inquiry were self-imposed by the scientists, and invisible to the public. Even the self-denying ordinances on doing harm, on publication of results, or on the actual pursuit of research—from Hippocrates to Leo Szilard to the Asilomar conferees—were of a very different kind from those now being discussed or proposed in forums far from the laboratory bench. Indeed, even in the laboratory most of the constraints listed above are today still largely passive and invisible rather than being the subject of conscious examination or having an explicit place in the teaching of young scientists. This is only one more example of the well-known resistance among scientists to self-consciousness in the study of actual scientific practice; but in this instance, it also slows the chances for convergence between the interest of scientists and their monitors.

The Endless Frontier Revisited

Having pointed out some difficulties which the "old credo" of science has in today's world, one must hasten to acknowledge again its immense power and usefulness. It has helped nurture a strong science/technology community in the United States, and elsewhere. It has helped us to fashion two activities of great strength. First, as the meetings and publications of the major scientific societies amply show, the quantity and quality of basic research in many fields is higher than at any time in history. The work of these scientists, largely but not exclusively done in the universities, is usually motivated and measured by the standards of "pure" or "basic" science rather than public need. It is the product of a largely autonomous, self-governing system, not directed by the calculus of risk and benefits. If there are other affected interests, most of those are placed at a distance, and the scientists are insulated from them. The hope for social utility as a by-product of one's discipline-oriented research may be in the background. But the ruling motto is that "truth must set its own agenda."

Moreover, some of us, myself included, continue to insist whenever we can that the very business of developing a rational and functional model of the universe through basic research is *itself* a major social goal for any civilized society, and fulfills a public need on that account. We point to the triumphs in molecular biology or biophysics, in cosmology or elementary particle "zoology," and in many other fields, and we say that whatever changes may be made, the quality of the product must not be put in danger. Of course, there are costs: severe specialization (although it is not an unmitigated evil: at the very least it helps to define the peer group and hence the peer review system, and in any case it is often the only way to do anything useful at all); reductionism (but it helps to select manageable problems); and the vague discomfort that the public, which is paying for all of this work, no longer knows what we are doing, since the conceptions are increasingly sophisticated and our patience with educating the public is not notably high.

To turn to the second mode of current excellence, there is another portion of the scientific/technological community that knows how to apply the basic scientific findings to a multitude of "public needs," needs that are articulated by a variety of constituencies, from the householder to the Joint Chiefs of Staff (and occasionally, merely by advertising agencies): the list includes vaccines, nuclear reactors, 10^4 industrial chemicals per year, computers, moonshots, new food plants, photovoltaic cells, insecticides, the cruise missile, not to speak of frozen dinners and shaving creme coming out of cans at the push of a button. The accent is more on problem-oriented development than on "truth"-oriented research, although there is often an interaction between truth and utility, as in materials sciences and medical research.

The great problem that has surfaced in this second mode is of course that such work, done on behalf of expected customers in a competitive market, provides at most only first-order solutions, into which are built sometimes awesome, second- and third-order problems. The larger the scale of the technology involved, the more likely are these second- and third-order problems to surface eventually. The reason for these by-products is partly that the complexity of the situation is almost by definition very great (hence, they are called "unforeseen" new problems), partly that the training of even the applied scientists is generally not geared to such sober and, on the whole, boring business as technology self-assessment; and, above all, that the client—the state, the industry—is usually in a hurry, is reluctant to pay for more than first-order solutions, and until recently has made little attempt to get advice on hidden flaws until they have become obvious, and therefore less manageable.

This dichotomy of styles of research and development, the separation between them, and the unfavorable by-products, are the results of a long development of science and technology in the Western world. But at least as far as the United States is concerned the most formative part of that history has been the most recent period—specifically, the conscious choices made by the politically most powerful segments of the scientific community when the outlines of the present system of science support were set up at the end of World War II. That system was the result of a battle between the ultimately victorious forces, led by Vannevar Bush, the head of the Office of Scientific Research and Development, on the one hand, and the forces identified with the New Deal Senator Harley Kilgore on the other.

The issue between them was really an old one: What should be the main thrust of science and hence the main justification for its support in our democracy? As Daniel Kevles has noted in his recent book, *The Physicists,* since its start in the United States, the debate has been polarized along the same axis, one that may be quite roughly characterized as knowledge for its own sake versus social usefulness.

To be sure, one must not only look at the extremes of these alternatives, for the progress of both "pure" and of technological application has become more and more strongly intertwined. The pre-World War II debates between Hogben, Crowther, and Bernal on the one hand, who argued for the primacy of the function of science in the promotion of human welfare, and, on the other hand, their opponents such as Polanyi, who favored the pursuit of science dissociated from any conscious connection with practical benefits, created so sharp a dichotomy that the arguments had to be ultimately inconclusive. Today, we know well that advances in the most recondite and speculative branches, such as cosmology, can depend on data obtained with space- and computer-age technology, while the most mundane electronic gadgets trace their lineage to the early papers on pure quantum physics by Planck, Einstein, and Bohr.

Nevertheless, it matters greatly where the center of gravity is placed in a large national institution for the support of science, and how large a spectrum is encouraged. Kilgore's declared aim was to set up a national research and development foundation which would assure that at least a major part of federally supported science research be linked to a social purpose, and that progress in science be planned to some extent by such an agency. Vannevar Bush and his colleagues, however, were opposed to this model, and were against such notions of Kilgore as a major revision of patent rights flowing from government-sponsored research, or geographical criteria for the distribution of funds, or the inclusion of support for the social sciences in a super agency for research. They also had to worry about attacks from the other wing. For at that end of the political spectrum, the opinions about Bush's own proposals were, as Bush noted in his autobiography, *Pieces of the Action* (p. 64), that "we were inviting federal control of the colleges and universities, and of industry for that matter, that this was an entering wedge for some form of socialistic state, that the independence which has made this country vigorous was endangered." Bush himself had similar fears about Kilgore's ideas; as Bush's associate Homer W. Smith put it, the scientist needs "the intellectual and physical freedom to work on whatever he damn-well pleases."

One can understand why, particularly at that time, the working scientists may have felt keenly that way. World War II had just ended, and the scientists were concerned that the armed services were going to keep their hands on the direction of research. Even the *New York Times*, under the heading "Science and the Bomb," declared editorially on the day after the bomb devasted Hiroshima (in its August 7, 1945, edition) that the scientists had better shape up and learn a lesson from the events:

> University professors, who are opposed to organizing, planning and directing research after the manner of industrial laboratories because in their opinion fundamental research is based on 'curiosity' and because great scientific minds must be left to themselves, have something to think about. A most important piece of research was conducted on behalf of the Army by precisely the means adopted in industrial laboratories. And the result? An invention is given to the world in three years which it would have taken perhaps half a century to develop if we had to rely on prima donna research scientists to work alone. The internal logical necessity of atomic physics and the war led to the bomb. A problem was stated. It was solved by teamwork, by planning, by competent direction, and not by a mere desire to satisfy curiosity.

Such omens, and the memory of scientists having to buckle under in totalitarian countries, were a source of considerable concern to many American scientists. While Kilgore wanted the best science required for the national needs of the United States, Bush thought it safest to ask for "the best science, period." In a free society, he hoped, it would automatically serve the national need. The model for that to happen was at hand, and Bush confessed freely that he used it to advantage in lobbying for his planned agency:

> There were some on Capitol Hill who felt that the real need of the post-war effort would be support of inventors and gadgeteers, and to whom science meant just that. When talking matters over with some of these, it was well to avoid the word "fundamental" and to use "basic" instead. For it was easy to make clear that the work of scientists for two generations, work that had been regarded by many as interesting but hardly of real impact on a practical existence, had been basic to the production of a bomb that had ended a war" (op. cit., p. 65).

Other institutions—industry, the Office of Naval Research, the planned Atomic Energy Commission—could be counted on to support research "along the lines of their special interests," but this would not be true for "basic research, fundamental research" (ibid.). This gap was to be filled by the National Science Foundation. Without a dedicated agency of this kind, Bush held, "a perverse law governing research" would assert itself: "Under the pressure for immediate results, and unless deliberate policies are set up to guard against this, *applied research invariably drives out pure.* This moral is clear: it is pure research which deserves and requires special protection and specially assured support" (*Science, the Endless Frontier*, p. xxvi).

Thus, the public credibility of Bush and his fellow scientists in their drive to invent institutions and obtain previously undreamt of financial support for basic research came at the time not from the persuasiveness of their own basic research performance—that work had been laid aside during the war years—but rather from the superb job that had been done in applied science and development during the war. And although the irony is clear now, it was then quite natural that the top scientists, turned war-time engineers, whose work had started the nuclear reactor wastes flowing into the fragile tanks at Hanford, Washington, and elsewhere, now wanted to hurry back to "best science." They did not even stay to ask what important scientific puzzles would have to be solved before the nuclear wastes could be safely disposed of. In fact, a full seventeen years had to elapse from the first design, around 1940, of nuclear reactors by some of the best scientists of the time (Fermi, Szilard, and their colleagues) until the publication of the first major study of high-level waste management.

Ironically, too, Vannevar Bush obtained his mandate to write what became the ground plan for the National Science Foundation from Franklin D. Roosevelt, who was personally most attracted to the *applied* side of the plan. In his original letter to Bush, Roosevelt asked for a way to put to use "the information, the techniques, and the research experience" developed during the war in order to wage a "war of science against disease," to "aid research activities by public and private organizations," and to devise ways of "discovering and developing scientific talent in American youth" (*Science, the Endless Frontier*, pp. 3–4).

In his response, the seminal report, *Science, the Endless Frontier*, Bush followed that rhetoric closely. He gave low visibility in print to the pursuit of pure science for its own sake, relying heavily instead on the tenuous promise that somehow science will be of social use. It is, to this day, a splendid document. It captures the well-deserved, utopian hopes at the end of a terrible war, and in many details was visionary, necessary, and right. But it does strike us now forcefully that in the report no mechanism was specified or encouraged for the benign spinoff-effects of basic science. It was not specified how the central alchemy would occur that would yield the promised long-range benefits. And of course no mention was then made of the other, darker side of the coin, the possibility of negative consequences, not to speak of the need for impact statements and the like. On the contrary, the very first paragraph of Bush's report starts with this clarion call:

> Scientific progress is essential.
>
> Progress in the war against disease depends upon a flow of new scientific knowledge. New products, new industries, and more jobs require continuous addition to knowledge of the laws of nature, and the application of that knowledge to

> practical purposes. Similarly, our defense against aggression demands new knowledge so that we can develop new and improved weapons. This essential, new knowledge can be obtained only through basic scientific research.
>
> Science can be effective in the national welfare only as a member of a team, whether the conditions be peace or war. But without scientific progress no amount of achievement in other directions can insure our health, prosperity, and security as a nation in the modern world (ibid., p. 1).*

In itself, this program and this promise, and even this language, are not at all that new. For some centuries, the announcements of great projects of science have had a remarkably similar ring. For example, the preamble of the Act establishing the American Academy of Arts and Sciences in 1780 begins with the presupposition of beneficence:

> As the Arts and Sciences [effectively, a range of subjects from cosmology to medicine and horticulture] are the foundation and support of agriculture, manufactures, and commerce; as they are necessary to the wealth, peace, independence, and happiness of a people; as they essentially promote the honor and dignity of the government which patronizes them; and as they are most effectively cultivated and diffused through a State by the forming and incorporating of men of genius and learning into public societies for these beneficial purposes, Be it therefore enacted. . . .

Almost exactly two hundred years later, President Carter wrote in his covering letter (2 August 1978) for the Tenth Annual Report of the National Science Board that "Basic research is the foundation upon which many of our Nation's technological achievements have been built," and the news release of the National Science Foundation, accompanying the publication of his letter, put it more simply still: "Basic research is useful."

The promise of beneficence emerging from such documents has been routine, and (at least until lately) almost universally accepted as plausible. In Vannevar Bush's plan, the lack of specificity—how disease, ignorance, and unemployment would be conquered if basic science is supported on a large scale—also had a number of good reasons. Not the least may be that the very image of science as the "endless frontier" avoids the need for operational details of this sort. In his book, *The American Character* (1956), D. W. Brogan made the astute observation that American and English thought differ in the connotations of the idea of the "frontier":

> The frontier in English speech is a defined barrier between two organized states; in American it is a vague, broad, fluctuating region on one side of which is a stable, settled, comparatively old society, and on the other, empty land, a few savages, unknown opportunities, unknown risks. American history has been a matter of eliminating that debatable area between the empty land and the settled land, between the desert and the town. This elimination has now been completed, but it is too early, yet, for the centuries-old habits to have changed and much too early for the attitude of mind bred by this incessant social process to have lost its power.

The implication of the beneficence, unforeseeable in detail but inevitable, of the effects trickling down from basic science to useful application continues to

*One more irony is that when the National Science Foundation was set up as the direct result of Bush's campaign, its mandate *excluded* the "war against disease," the support of medicine and its related sciences that may have been of greatest personal interest to Roosevelt, as well as all civilian research on weapons.

be held before the public to this day. An effect of this sort of course exists. But it is really still little understood, is difficult even to reconstruct historically, and is now losing rather than gaining institutional support as industry is increasingly turning away from the support of basic research. In good part for these reasons, few scientists would want the agencies that are now committed to support basic research to change that commitment in any major way. Yet the public has by no means forgotten about the implied IOU's; and as the combined bill for both elements of the dichotomy, "pure science" and "useful science," has greatly increased over the years, and as the bridge from one to the other has not become clearer, people from Senator Proxmire on to the right and left have become more critical. In addition, Don K. Price correctly remarked, in his article in this volume (p. 85), "The danger of political constraints on science now comes not so much because politicians disapprove of the methods of science, but because they take seriously what some scientists tell them about the way in which scientific discovery leads to practical benefits. Once they believe that, it is inevitable that they should try to control practical outcomes by anticipating the effects of research, and manipulating it in one way or another." It may therefore be the final irony that the implied but difficult-to-deliver promises in the Bush report, on the basis of which the dichotomy was achieved and the pure-science ideal triumphed, became a factor in the current assertion of external constraints on science.

The Precariousness of the New Bargain

Even as the old institutional arrangements for debate and agreement now seem inadequate, so does the old bargain between science and society itself. The generally accepted model of the linkage, in typical cases, involved a barter of an odd sort. That is, as in the Bush report, society could expect in the long run to get certain though not well-defined material benefits from the work of researchers in science and technology, and those individuals in turn expected to benefit by receiving some moderate financial support and considerable administrative freedom from society. In that barter the two sides were in most instances distant and rather unengaged. Each gave up to the other things that it could easily afford to yield, or that it did not particularly treasure. Each side had no great incentive to understand the other, and could be satisfied with the gains it was eventually receiving for itself. The system worked well, and was moderately well understood.

But the model of interaction that has been developing lately is rather different, and very much in need of study. (As Harvey Brooks says in this volume, we do not even know how science and development are connected with social policy: which drives which?) The new model involves (again in barest outline) the giving up by each side of substantial, treasured items, and the exchange requires much more mutual involvement and understanding. Each side now has to barter away some of its autonomy, and each side expects to obtain substantial, monitored benefits and assurances. Much of the argument on where the limits of scientific inquiry should be is a by-product of this encounter.

The mood I see developing in this encounter corresponds to the impression I gained at the end of the series of meetings of the faculty seminar at MIT at which most of the essays in this volume were discussed in draft form. If there was a consensus, it was, in the words of one of the authors, that just as we are wary of the "slippery slope" in biomedical ethics, so we must resist slipping inadvertently into increased controls over fundamental science, since such controls can easily lead to abuses of their own.

The image shared by most scholars there was that we are now dealing with *two* slippery slopes, joined at the top by a razor's edge upon which we are precariously balanced. There will thus be hard bargaining on such questions as these: *Who* will control what, and at what level? (At the level of the institution, of the individual researcher, of funding?) How can "public participation" be arranged without clashing with the very meaning of science as a consensual activity among trained specialists? (In the old bargain, the inherent contradiction between science and democracy did not have to be faced. Now, it does.) When today's largely unwritten code of ethics, maintained by peer pressure, is turned in for a set of congressionally developed guidelines, how can enough room be left in them for personal choices that will need to be made as conditions and knowledge change? (This is analogous to obtaining "variances.") Or will such choices be smothered by blanket regulations, formulated by bureaucratic entities that have considerable inertia? In that ominous event, how can the freedom and momentum be preserved upon which any imaginative group must be able to draw to do its best?

Need for New Institutional Forms

The fact that the questions posed in the last paragraph have today no persuasive answers indicates that there is a real paucity of institutional forms for dealing with them. As one reads proceedings of conferences or congressional hearings in which representatives of the scientific establishment and of the public struggle with the new problems of accountability, one is impressed how often these old institutional forms for mutual negotiation really have become inadequate for the complex task that is required. Instead of the model of an interdisciplinary investigating team in which the members assist one another in a common task from which all would benefit, one finds all too often the model of unilateral pronouncements inherited from the classroom, or of adversary proceedings as learned in the courtroom.

Institutional forms are slow to grow, and until enough credible ones have been designed and tested, the public may well continue to push for ad hoc solutions that may not at all appeal to the scientists. It is a very hopeful sign of their new maturity that the scientists, too, are in fact hard at work fashioning new institutional devices and experiments. A brief list would include the role scientists have played in the Asilomar conference, the various citizens advisory panels, federal funding of scientists to work with public interest groups or with projects on the public understanding of science, the revision of codes of ethics by some professional societies (to make such codes something more meaningful than the traditional, rather self-serving documents), the federally-funded courses concerned with ethical and value impacts of science and technology, the sections within professional scientific societies such as the American Physical Society (its Panel on Public Affairs, its Forum on Physics and Society, its summer study groups on nuclear waste management and other problems, its congressional fellowships, etc.). A glance at the table of contents of the journal *Science* and of the meetings of its parent organization, the American Association for the Advancement of Science, will show how profound the changes are which have taken place over the past dozen years in the attention of that large group. And of course there are many other signs—the emergence of the Society for the Social Study of Science, the increase in modern case studies in the field of the history and sociology of science, and last but not least, academic programs such as the new College for Science, Technology and Society at MIT, one of whose activities was the sponsorship of the meetings which resulted in this very volume.

To illustrate the novelty and, in the face of controversy, daring of such activities in which scientists themselves are participating, I will single out only one example. It concerns the discussion within research institutions on the question of impact statements. In seeking funding or the acceptance of proper research procedures, scientists and engineers have for some time now accepted their liability for submitting impact statements dealing with environmental impact, financial responsibility, manpower policy, affirmative action, and the like. Nobody has liked such bureaucratic impositions, and they can become detrimental to the execution of the work when they are too burdensome; but at least they are now being routinely attended to. Lately, there have been some discussions that sooner or later ethical impact or values impact statements may be added in certain areas of inquiry, and that therefore scientists and engineers should begin to become familiar with such conceptions before they are imposed unilaterally from without.

To a degree, this was the message of the Bauman amendment in 1976 by which the Congress would have reserved for itself the right to reexamine each grant made by the National Science Foundation after peer review. Depending on the effect or impact this research might be thought to have, in the opinion of any Congressman, an embargo could have been put on a particular grant after it was made. As part of a similarly motivated action, in the authorization bill for appropriations for the National Science Foundation issued in March 1977 the Congress directed the NSF to provide for the protection of students in try-out classes that use precollege educational materials. The aim was to make sure that "interested school officials" could make certain that steps had been taken "to insure that students will not be placed 'at risk' with respect to their psychological, mental, and emotional well-being" while serving as human subjects in these educational developments.

The idea of asking researchers for a statement of the prospective impact of their research has been circulating quietly. Thus in March, 1974, a faculty committee was set up at Massachusetts Institute of Technology to see how faculty members could make a more thorough and self-conscious study of "the impact of MIT research insofar as that research has influence on matters such as the physical environment, the economy, national security, and other important social concerns." One function of the committee was to develop methods "to assist individual investigators in the preparation of impact statements." In making its report to the faculty, the committee—one containing such widely respected members as Frank Press and John Deutch—said "An honest effort to estimate the plausible consequences of his own research is as properly to be expected from a researcher as now he is expected to estimate the cost of carrying out his work, and is in the long run perhaps more significant."

When it came to the vote, the MIT faculty felt that the time was not ripe for this experiment. However, it was made clear in the discussion that the issue should not be dropped, and that a more specific proposal should be developed for future action. I expect we shall hear more about this and similar ideas, aiming to fashion new institutions for dealing with the problems of limits that arise when a scientific pursuit or engineering project has the potential of encroaching on other widely held social values—and to do so before the problem of setting limits has entirely escaped from the hands of the scientists themselves.

Ideology of Limits

Another main force driving the events which preoccupy the authors in this volume is the change now underway in the developed countries of the West, away from the old "ideology of progress," and toward a new "ideology of limits" that goes much beyond limits to scientific inquiry. One of the lessons of the sensible decisions not to build the SST in the United States and not to go to a plutonium economy is that a nation's leadership now can invoke an almost unprecedented self-denial of technology, a turning away from the "can do means must do" imperative. To a certain degree, an embargo on some aspect of science can be characterized as being a call for a stop to the pursuit of knowledge of *how to* (rather than knowledge of *why*), and is therefore closely similar to the self-denial of these new technologies.

This development fits in with the general new awareness of the existence or necessity for *limits* (a word that may yet come to characterize a main preoccupation of the 1970s)—limits to natural resources, in particular, energy supplies; limits to the elasticity of the environment to respond beneficently to the encroachment of the man-made world; limits to food supplies; limits to population; limits to the exercise of power (including presidential power); and, within science itself, limits to growth of its manpower and institutions.

The pervasiveness and reality of this new climate of opinion is being more widely studied. Thus a major report made for the Joint Economic Committee of the Congress, entitled *Limits of Growth*, volume 7 of *U.S. Economic Growth from 1976 to 1986: Prospects, Problems, and Patterns* recently specified the existence of a number of popular new "emphases and trends" that are expected to influence growth during the next decade. These include the continued rise of risk aversion, concern with health and environmental protection, and antitechnological and anti-industrial attitudes. It is clear that for a substantial part of the population of the West, the image of this century has been changing from the era of frontiers waiting to be exploited, to the era of the globe as a crowded lifeboat.

New Conception of Progress

The last and perhaps most decisive component in the current shaping of events has to do with the rise of a new conception of progress in science itself. Here too a substantial change in Zeitgeist has been taking place, and it surely reinforces the effectiveness of the limits-to-inquiry movement.

To be sure, the old notions of the purity and objectivity of science and of its claims to truth in some absolute sense have been under attack for a long time. But this has been largely a philosophical debate, with little effect on most research scientists whom such epistemological—perhaps to some degree even quasi-theological developments—reach with a long time delay and with few credentials. Thus scientists see themselves engaged in an enterprise, in which they consider progress and cumulation to be perfectly well identifiable, and in which the chief marks of progress are still taken to be the classical ones: greater inclusiveness of separate subject matter, and greater parsimony or restrictiveness of separate fundamental terms.

Until not too long ago, the popular conception of science as inexorably "progressing" had been a component of thought even in the historiography of science. George Sarton went so far as to assert in 1936, "The history of science is the only history which can illustrate the progress of mankind. In fact, progress has no definite and unquestionable meaning in other fields than the field of science." But just in the last few years these progressivist assumptions have come under a variety of attacks from scholars in the history and philosophy of science. Among those, I select two who agree with each other on little, except on the rejection of the old notion of progress. Each has a wide following (and their published opinions, excerpted below, only indicate the flavor of the writings by others whose versions lack the acknowledged subtleties of the more fully developed arguments).

The first school, headed by the historian of science T. S. Kuhn, regards the notion of scientific progress as a self-fulfilling definition, therefore essentially a tautology. He asks, "Viewed from within any single community . . . the result of successful creative work *is* progress. How could it possibly be anything else?" The upshot is, we are told, that we may "have to relinquish the notion" that the large changes which science occasionally undergoes could carry scientists and those who learn from them "closer and closer to the truth." An implicit consequence is not far away: when it is widely believed that scientific progress is defined only by whatever the scientists are doing anyway, then limiting some specific work in the presumed interest of other, more urgent human values will seem a far less intolerable intervention.

The other school, headed until his recent death by the philosopher of science Imre Lakatos, quite explicitly aims to set up, as Lakatos put it, "universal criteria" for distinguishing "progressive" research programs from those he regarded as "degenerating." He wrote that he did so partly in order to help editors of scientific journals to "refuse to publish" such papers, and research foundations to refuse support for such work. (He warned, "Contemporary elementary particle physics and environmentalist theories of intelligence might turn out not to meet these criteria. In such cases, philosophy of science attempts to overrule the apologetic efforts of degenerating programs.") To help the layman to decide what scientific project may or may not be pursued or supported, this school announces finally that it will "lay down statute law of rational appraisal which can direct a lay jury in passing judgment."

This is not to imply that public officials who want to stop scientific research or engineering projects within their cities are motivated by reading books in the history or philosophy of science. Rather, the theories of progress in those books seem to me to be a part of a current of thought in which working scientists who still have the earlier, progressivist notions now appear, to some, as dangerous meddlers. We have in fact entered a period where old and new theories of progress in science are vigorously competing—in the mind of the public, among those engaged in the study of nature, and among scholars who study science as an activity.

This current, like the other forces I have cited, is not amenable to sudden change. The momentum of the debate is large, and the divisions on fundamental issues are deep. It is therefore not reasonable to expect quick solutions. A better hope—as has been attempted here—is to gather a greater variety of interested parties to see what we may learn from one another, and thereby improve the chances for wiser action and accommodation. For the issue is still the same one which Watson Davis, with a different aim, identified when he said over forty years ago: "The most important problem before the scientific world today is not the cure of cancer, the discovery of a new source of energy, or any other specific achievement. It is: How can science maintain its freedom, *and* . . . help preserve a peaceful and effective civilization?" As this volume indicates, scientists, in larger numbers than ever before, are wrestling with both parts of the question, knowing perhaps that if they wish to answer one of these, they must answer both together.

Notes on Contributors

DAVID BALTIMORE, born in 1938 in New York City, is American Cancer Society Professor of Microbiology at the Massachusetts Institute of Technology. His publications include articles on nucleic acids, cancer, and viruses. In 1975 he received the Nobel Prize in Physiology or Medicine.

SISSELA BOK, born in 1934 in Stockholm, Sweden, is lecturer at the Harvard Medical School. She is a member of the Ethics Advisory Board of the Department of Health, Education, and Welfare, and is author of the forthcoming *Lying: Moral Choice in Public and Private Life*.

HARVEY BROOKS, born in 1915 in Cleveland, Ohio, is Benjamin Peirce Professor of Technology and Public Policy at Harvard University. From 1971 to 1976 he was president of the American Academy of Arts and Sciences. Among his publications are *The Government of Science* (1968), *Controversies and Decisions* (1976), and articles in *When Values Conflict* (1976).

BARBARA J. CULLITON, born in 1943 in Buffalo, New York, is the editor of "News and Comment" for *Science* magazine. She has written numerous journal and magazine articles.

LOREN R. GRAHAM, born in 1933 in Hymera, Indiana, is professor of history at Columbia University. He is author of *The Soviet Academy of Sciences and the Communist Party* (1967), and *Science and Philosophy in the Soviet Union* (1972). In 1976–77 he held a Rockefeller Foundation Humanities Fellowship at the Program for Science and International Affairs, the Department of the History of Science, and the Russian Research Center at Harvard University.

GERALD HOLTON, born in 1922, is Mallinckrodt Professor of Physics and professor of the history of science at Harvard University, and is currently visiting professor at the Massachusetts Institute of Technology. From 1957 to 1963, he was editor of the American Academy of Arts and Sciences and, until 1961, editor of *Daedalus*. Among his publications are *Thematic Origins of Scientific Thought: Kepler to Einstein* (1973), *Introduction to Concepts and Theories in Physical Science* (1952, 1973) and *The Scientific Imagination: Case Studies* (1978).

PETER BARTON HUTT, born in 1934 in Buffalo, New York, is an attorney in Washington, D.C. Formerly general counsel of the Food and Drug Administration, he has been a member of the National Academy of Sciences Committee on National Needs for Biomedical and Behavioral Research Personnel, of the National Institutes of Health Committee to Review Guidelines for Research on Recombinant DNA Molecules, and of the Institute of Medicine Health Science Policy Advisory Committee.

LEO MARX, born in 1919 in New York City, is Kenan Professor of American Cultural History at the Massachusetts Institute of Technology. He is author of *The Machine in the Garden: Technology and the Pastoral Ideal in America* (1964), and has contributed articles to scientific and literary journals.

WALTER P. METZGER, born in 1922 in New York City, is professor of history at Columbia University. His publications include *Academic Freedom in the Age of the University* (1955), and, with Paul Goodman and Sanford J. Kadish, *Freedom and Order in the Uni-*

versity (1967). He served as editor for the collection, *Academic Professional in the Western World* (1976).

ROBERT S. MORISON, born in 1906 in Milwaukee, Wisconsin, is currently visiting professor at the Massachusetts Institute of Technology. He has served as director of medical and natural sciences at the Rockefeller Foundation, and as director of the Division of Biological Sciences at Cornell University. He is author of *Scientist* (1964), editor of *Contemporary University: U.S.A.* (1966), and has written many articles on science education in developing countries and in American universities.

DOROTHY NELKIN, born in 1933 in Boston, Massachusetts, is professor of city and regional planning and of the Program in Science, Technology, and Society at Cornell University. Her recent publications include *University and Military Research* (1972), *Methadone Maintenance—A Technological Fix* (1973), *Jetport: Boston Airport Controversy* (1975), *Science Textbook Controversies* (1977), and *Technological Decisions and Democracy* (1977).

DON K. PRICE, born in 1910 in Middlesboro, Kentucky, is professor of government at Harvard University. He has served as president of the American Association for the Advancement of Science, as dean of the John Fitzgerald Kennedy School of Government at Harvard University, and is chairman of the Twentieth Century Fund. His publications include *Government and Science* (1954), and *The Scientific Estate* (1965).

ROBERT L. SINSHEIMER, born in 1920 in Washington, D.C., is chancellor of the University of California at Santa Cruz. Formerly chairman of the Department of Biology at the California Institute of Technology, his publications include numerous articles on nucleic acids, viruses, and the societal implications of modern biology.

JUDITH P. SWAZEY, born in 1939 in Bronxville, New York, is professor of sociomedical sciences at the Boston University School of Medicine. Her recent books include *Reflexes and Motor Integration: The Development of Sherrington's Integrative Action Concept* (1969), *Chlorpromazine in Psychiatry: A Study of Therapeutic Innovation* (1974), and, with Renée C. Fox, *The Courage to Fail: A Social View of Organ Transplants and Hemodialysis* (1974).

LYNN WHITE, JR., born in 1907 in San Francisco, California, is University Professor of History emeritus at the University of California, Los Angeles. He is author of *Latin Monasticism in Norman Sicily* (1938, 1968), *Medieval Technology and Social Change* (1962), *Machina ex Deo* (1968), and the forthcoming *Medieval Religion and Technology: Collected Essays*.

Index